SR-71

The Complete Illustrated History of the Blackbird, The World's Highest, Fastest Plane

Col. Richard H. Graham, USAF (Ret.)

ZENITH PRESS

Brimming with creative inspiration, how-to projects, and useful information to enrich your everyday life, Quarto Knows is a favourite destination for those pursuing their interests and passions. Visit our site and dig deeper with our books into your area of interest: Quarto Creates, Quarto Cooks, Quarto Homes, Quarto Lives, Quarto Drives, Quarto Explores, Quarto Gifts, or Quarto Kids.

First published in 2013 by Zenith Press. This edition published 2017 by Zenith Press, an imprint of The Quarto Group, 401 Second Avenue North, Suite 310, Minneapolis, MN 55401 USA.
T (612) 344-8100 F (612) 344-8692 www.QuartoKnows.com

10 9 8 7 6 5 4 3 2 1

ISBN: 978-0-7603-5448-3

Library of Congress Cataloging-in-Publication Data for the hardcover edition:

Graham, Richard H., 1942–
    SR-71 : the complete illustrated history of the Blackbird, the world's highest, fastest plane /
Col. Richard H. Graham, USAF (Ret.).
        pages cm
Includes bibliographical references and index.
ISBN 978-0-7603-4327-2 (hbk.)
1. SR-71 Blackbird (Jet reconnaissance plane)—History. I. Title. II. Title: Complete illustrated history of the Blackbird, the world's highest, fastest plane.
UG1242.R4G733 2013
358.4'583—dc23
                        2012044644

Acquiring Editor: Scott Pearson
Art Director & Cover Design: James Kegley
Designer: Diana Boger
Poster Illustration: Mark Balfe

Page one: *Lockheed Martin*
Title page: *Lockheed Martin*
Back cover: *Tech. Sgt. Michael Haggerty/USAF*

Printed in China

For all the military and civilian personnel who worked on the SR-71.

Contents

Foreword

This latest book in Col. Rich Graham's series on the SR-71 provides comprehensive, articulate, and well-illustrated coverage of the program's entire history. The recent declassification of many details on the development and operations of the early A-12 version of the aircraft under the CIA's Oxcart program help make this broader understanding possible.

My earliest inkling that this kind of aircraft existed came before I entered the intelligence profession. As one of the first tourists allowed to travel independently in the U.S.S.R., I rode a Russian passenger ship into the Far East portal at Nakhodka harbor, just across the peninsula from the Vladivostok Soviet naval complex, on a bright, sunny afternoon in late 1972. As we approached the dock, I spotted a very high contrail moving toward us from the south, parallel to the coast, at a jaw-dropping rate of speed. The aircraft generating it was so high that nothing could be seen but a very tiny speck a long distance in front of it. In less than half a minute it disappeared to the north—but not before what were clearly two missile contrails rose vertically to intercept and then fell away, short of it and far behind. I later learned that there had indeed been an SR-71 mission in that theater, on that day. The experience was unforgettable, like most things associated with the Blackbirds.

The SR-71 family of aircraft met a clear intelligence need at an important time in U.S. history. Together, they survived almost as long as the Cold War. They rightly have a legendary place in the annals of aviation history. All of the manned aircraft speed and altitude records of these aircraft are still unsurpassed after more than fifty years since the first A-12 flight. The pioneering technologies invented for these aircraft and their sensors also paved the way for much of our space and military-stealth capabilities today.

Everything about the SR-71 family of aircraft and the unique individuals who developed its design variants, tested them, flew them, managed their highly classified missions, or supported those missions merits superlatives. There are few better examples of the American ability to quickly address pressing needs by employing expert knowledge, dedicated teamwork, flexible management, and continual, adaptive innovation.

Though I was not directly associated with either the Oxcart or SR-71 programs during my thirty-eight years of federal service in CIA and DIA in various intelligence positions, in many different job contexts I developed a thorough appreciation for these aircraft, their capabilities, and the dedicated professionals who designed, developed, piloted, and supported them.

In mid-career I was able to task the SR-71 several times to support advanced sensor evaluation projects that I helped manage. The response satisfied our data needs and image requirements quickly and precisely, answering crucial questions that eliminated development guesswork and reduced costs for future reconnaissance capabilities.

The SR-71 earned a solid reputation for reliably providing good results. More than a few intelligence successes (some of which are still classified) were achieved through the use of these extraordinary reconnaissance aircraft. They could appear on the scene unwarned, gather synoptic sensor data over a broad area, and leave so quickly there was no time for the target nation to prepare or react effectively. In the intelligence community, there was such respect for its many uses that there are still SR-71 models on desks of intelligence professionals today.

In recent years I have been fortunate to become closely acquainted with some of the key men involved in these programs as I entered on a personal, but work-related quest to help them recount their remarkable teamwork experiences to government employees, congressional staffs, and the public. I am convinced that the exceptional work they did demands that their

stories be told and retold at every opportunity. Doing so inspires our national leaders and intelligence and service professionals, as well as the public. It also may encourage similarly risky technological investments in other important areas.

I met Rich early in this process, and soon convinced him to recount his exceptional SR-71 experiences before a very receptive group of industry and intelligence professionals. We have since participated in related historical events (such as the dedication of the A-12 memorial at CIA) and worked together closely to facilitate the 2010 Oxcart Legacy Tour sessions at Washington, D.C., government and museum venues. We became good friends, having a common goal in the process. I am most pleased to see that in this book Rich is relating the *whole* story of these programs in a readable way for a larger public audience to enjoy.

I have often visited the SR-71 and A-12 aircraft now on display at the Smithsonian's Udvar-Hazy Air and Space Museum, the CIA, and the USAF museums in Dayton, Ohio, and at Beale AFB, California. Each meeting with one of these Blackbirds is almost mystical. The feeling that immediately comes upon one at first glimpse is that you are in proximity to greatness. After you read this excellent book, I hope that you, too, will make the effort to see these timeless aircraft up close in some of their museum locations.

I heartily commend this enjoyable book to you. It has something for everyone, from those who admire the SR-71 legend, to pilots of other aircraft, who will truly appreciate Rich's personal stories about the extraordinary complexities of flying this amazing machine. In my view, this is his best work yet.

—Peter W. DeForth
Retired Senior Scientific
Intelligence Officer, CIA

Acknowledgments

I would like to acknowledge the contributions of various friends, coworkers, and former colleagues who enriched this book with their perspectives and recollections. My good friend Jay Miller, who I regard as the greatest aviation authority in the United States, has allowed me to use previously uncovered "logs" kept by Kelly Johnson. They add tremendous personal insight into early development of the various Blackbird programs. Jay's documentation in his book, *Lockheed's Skunk Works: The First Fifty Years,* is unmatched in terms of details and added immeasurably to this book.

I met Peter DeForth many years ago while he was searching for a specific SR-71 print. He ended up with the print, and I ended up having my SR-71 books for sale in the CIA's gift shop! Peter has graciously opened up many doors and special events for me within the Washington, D.C., intelligence community. He knows the "product" end of the SR-71 better than anyone else I have ever met. Peter has been kind enough to write the foreword to this book and I am deeply indebted to him for doing so.

Another great aviation author that I have the highest respect for is Paul Crickmore, who lives in England. I've known Paul since he was working in London Air Traffic Control Center at RAF West Drayton in the early 1980s and stay in touch with him even today. His detailed knowledge of the early beginnings of the Blackbird program was extremely useful. His most recent book on the SR-71, *Lockheed Blackbird: Beyond the Secret Missions* was invaluable as a resource to accurately portray the early days of the program.

I need to thank several SR-71 crew members for using their personal stories throughout this book. Ken Collins' story of applying to be a CIA pilot flying the A-12 aircraft is an important part of the SR-71 heritage. Test pilot Bill Weaver's tale of having an SR-71 disintegrate around him at Mach 3.2 is truly amazing. Bernie Smith and Frank Stampf allow the reader to have a firsthand account of what it meant to be a Habu when things were not going well. I am indebted to them for using their stories. Arnie Gunderson was our onsite Pratt & Whitney J-58 engine expert. Anything you wanted to know about the SR-71's engine and inlet, Arnie could break it down into basics that even dumb pilots could understand.

Tony Landis and Lockheed Martin's Kevin Robertson were instrumental in providing the bulk of the pictures. T. D. Barnes, Rick Poole, Mary K. Stein, Bill Gornik, Mac McKendree, B. C. Thomas, Doug Soifer, Don Emmons, Ed Bethart, Mike Hull, David Allison, Bill Flanagan, Ken Collins, Toni Hiley at the CIA, and Martin Lupton have been very gracious in providing additional pictures. Tom Alison, Bob Murphy, and Mike Relja provided detailed information that I could not have found otherwise. A huge thanks to all the above individuals for their support in making this book possible.

A final thanks to my wife, Pat, who proofread and edited the book to make me look good. She had the patience to let me pursue writing this book over the last year and encouraged me to keep going. I couldn't have done it without her support!

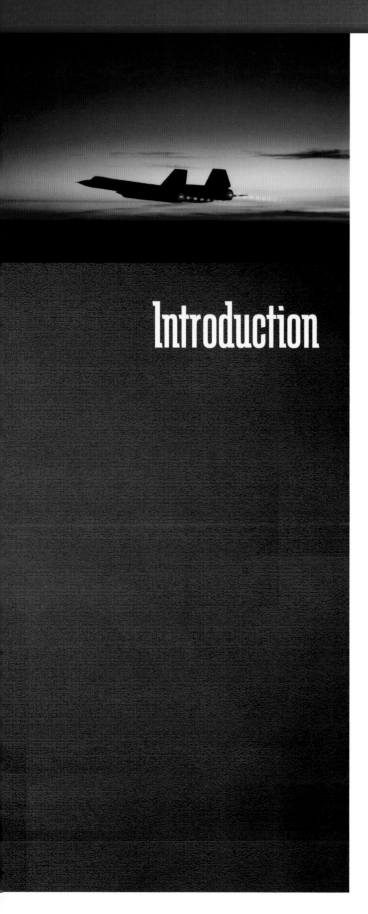

Introduction

Before the Blackbirds were even thought about, there was the U-2, a reconnaissance aircraft built specifically to gather intelligence on foreign countries. The chief architect and designer of the U-2 was Clarence L. "Kelly" Johnson, head of Lockheed's Advance Development Projects, better known as the Skunk Works. On 4 August 1955 the first U-2 took to the skies. Kelly knew that someday it would become vulnerable to enemy defenses, and, as early as 1958, he began to conceptualize a reconnaissance aircraft that would fly higher and over three times faster than the U-2.

Lockheed received approval on 28 August 1959 to proceed with the development of five A-12 aircraft, under the Central Intelligence Agency (CIA) code name Oxcart. On 26 April 1962, just twenty-two months after Kelly started building the first A-12 Blackbird, he had the plane flying with test pilot Lou Schalk at the controls. Lockheed produced a total of fifteen A-12 aircraft for the Oxcart program. The CIA pilots flew twenty-nine operational missions out of Okinawa, Japan, from 31 May 1967 to 6 May 1968, and the planes were then retired.

Two of the A-12 aircraft were converted to a two-seat configuration to carry the D-21 drone and were redesignated as the M-21. The drone was launched off the back of the M-21 at Mach 3 speeds, designed to penetrate deep into enemy territory and flew a preprogrammed course gathering intelligence from high altitude. At the completion of its mission, it would fly into international airspace and eject an intelligence package attached to a parachute, which was then snagged in mid-air by an Air Force cargo plane. The drone then self-destructed. Due to mixed success in obtaining good intelligence, the drone program was short-lived, and it was cancelled on 15 July 1971.

Another Blackbird variant, the YF-12, was designed to be a high-altitude, high-speed interceptor. They proved the technology and capability, but were soon canceled. Of the three built, two were lost in crashes. The remaining YF-12 is on display at the National Museum of the USAF in Dayton, Ohio.

The best known of the Blackbird family was the SR-71. Lockheed built twenty-nine SR-71A models, two SR-71B models, and one SR-71C model. The B and C models were trainer aircraft, designed and modified to teach aspiring SR-71 pilots. The first SR-71 flight took place on 22 December 1964, flown by Lockheed test pilot Bob Gilliland. Over the years, a total of eleven SR-71A models and 1 SR-71B were lost due to crashes. All remaining Blackbirds are on display at various aviation museums around the United States, with one on display in England.

Aviation fans often wonder what it was like to fly the SR-71 on secret Mach 3+ reconnaissance missions around the globe. My goal is to take you on a complete operational reconnaissance mission not only in words, but with pictures as well. The process of gathering worldwide intelligence on foreign countries with the SR-71 was both an art and science, refined over many years through searing Mach 3+ flight operations. The reader will have a better understanding of just how detailed and complex every SR-71 operational sortie was . . . from start to finish.

The Air Force's SR-71s flew 3,551 operational reconnaissance sorties from 9 March 1968 to 25 September 1989. Obviously, over twenty-two years of operational flying with the Blackbirds, many procedures changed and evolved with time, generally for the best. Don't be surprised if you talk to a former SR-71 crew member someday and he tells you, "That's not the way we did it when I was flying the plane."

My time associated with the SR-71 program spans 1974–1989, as a crew member, squadron commander, vice wing commander, and wing commander. Between 1982–1986 I worked in the Pentagon, and although not directly associated with the SR-71 program, I became very aware of the politics attempting to close down the Blackbird program. I remained close to all the former SR-71 crew members working in the Pentagon, who were attempting to keep the program viable during its latter years.

CHAPTER 1

Beginning of the A-12

The SR-71 story actually begins with the development of the U-2. In the early 1950s the United States was becoming increasingly concerned about a possible surprise attack from the Soviet Union's dreaded intercontinental ballistic missiles (ICBMs). The United States had no means of verifying their capabilities, which prompted the Air Force to set up a study group, looking into an aerial reconnaissance program. In mid-1952 the study group recommended every known technique be used and new ones developed, to increase U.S. intelligence over the Soviet Union by high-altitude photographic reconnaissance and other means.

Lockheed's Advance Development Project (ADP) boss at the time was Clarence "Kelly" Johnson. Hearing about this report, Kelly submitted a proposed aircraft known as the CL-282. After a short, but detailed, review, he received an official letter of rejection. Undaunted, Kelly decided to pursue funding for his high-altitude reconnaissance aircraft from other sources. He presented a refined U-2 aircraft design to a CIA study committee. On 24 November 1954, Kelly received the go-ahead to produce twenty aircraft at a total cost of $35 million. On 9 December the contract was signed between Lockheed and the CIA. The funding came from the CIA's secret Contingency Reserve Fund. Within days, Lockheed's ADP office had by default become a full-scale advanced design, engineering, and production facility. Kelly lived and died by his fourteen rules of management that suited his style perfectly and made the Skunk Works a true success in thinking outside the box (see appendices). The first test flight of the U-2 took place on 4 August 1955.

The CIA developed a cover story for the aircraft, stating that it had been developed as a high-altitude research tool for use by the National Advisory Committee for Aeronautics (NACA). Flying the U-2 over and around the Soviet Union took place out of various locations overseas by the CIA. Between 1957 and 1959 around thirty Soviet Union overflights took place. During 1958, the Soviets

Opposite: The current U-2S model carries state-of-the-art sensors able to collect intelligence in all weather and light conditions. The U-2 is employed during peacetime for intelligence planning and warning. Its wartime roles include battlefield surveillance, targeting, and battle damage assessment. *Lockheed Martin*

began to gather momentum in their efforts to develop an effective U-2 countermeasure. The SA-2, surface-to-air missile (SAM), was being developed with a warhead that had a kill pattern with a diameter of around 400 feet and posed a potential threat for the U-2. Even though the probability of a kill was low, for the first time the U-2 was forced into taking the new SA-2 launch sites into consideration by giving them a wide berth of up to thirty miles.

On 1 May 1960, CIA pilot Francis Gary Powers took off in a U-2 from a base near Peshawar, Pakistan. His reconnaissance targets were Sverdlovsk and Plesetsk, two major ICBM test sites in the Soviet Union. The Soviets were becoming more aggressive in their attempts to shoot down a U-2. During his photo run over Sverdlovsk, a salvo of numerous SA-2 missiles were fired at Powers' aircraft, and he subsequently ended up bailing out and being captured. Powers was tried as an American spy and put in a Soviet prison. On 10 February 1962, Powers was traded for the Soviet spy Rudolf Abel.

Several years prior to the Powers incident, Kelly Johnson and others believed that the U-2 would remain invulnerable for

no more than about two years after starting the overflights of Russia. In the fall of 1957 Kelly was asked to conduct an operations analysis to determine the probability of shooting down an aircraft and how it varied respectively with speed, altitude, and radar cross section (RCS). The analysis concluded that supersonic speed, coupled with the use of radar-attenuating materials and design characteristics, greatly reduced the chances of radar detection. Kelly and others

were impressed with the study's findings and agreed that further exploratory work should be done. Attention in the CIA now focused on building a vehicle that could fly at extremely high speeds and altitudes and would also incorporate the best available radar-attenuating capabilities.

Kelly spent virtually every available minute working on the advanced aircraft program, which he jokingly referred to as the "U-3." Between 1957 and 1959 Kelly and his team of engineers came up with several aircraft meeting the speed and altitude requirements for a reconnaissance platform. On 29 August 1959, Lockheed received official notification that it had won the competition with its A-12 design submission. On 4 September, Lockheed received $4.5 million by way of an advanced feasibility contract. The CIA's top secret project received the classified name: Oxcart.

Early on, radar cross section was determined by placing the aircraft on a pole in various attitudes and measuring its ability to be "seen" by precision radars. With the use of radar-absorbing ferrites and plastics on all the leading edges, the SR-71 has been described as "looking for a single-engine Piper Cub at 80,000 feet." *Lockheed Martin*

Former CIA pilot Ken Collins in front of an A-12. The yellow container on the ground supplies cooling air to the pressure suit during ground operations. Inflight cooling air is supplied from the aircraft air conditioning system.
Ken Collins

The sheer quality and quantity of imagery secured by the U–2 overflight program had become highly addictive to the U.S. intelligence community. However, the future of Oxcart was far from certain. The Powers shoot down proved to have a profound impact, as detailed in a top secret memo of 2 June 1960, written by General Andrew Goodpaster, military aide to President Eisenhower:

I spoke to the President early this week about the question of whether work should go forward on the successor to the U–2. After considering the matter, he said he was inclined to think it should go forward, on low priority, as a high performance reconnaissance plane for the Air Force in time of war. I suggested it might be useful for Mr. Allen Dulles, Mr. Gates and Mr. Stans to get together to consider the matter, and he agreed. He said he did not think the project should now be pushed at top priority. In fact, they might come to the conclusion that it would be best to get out of it if we could. Alternatively, they may feel that we have so much invested in it that we should capitalize on this through carrying it forward.

Funding allocated for Oxcart by the CIA also included developmental funding for the Pratt & Whitney J-58 engine. Pratt & Whitney had been involved with Oxcart from the very beginning. Their J-58 had been sponsored originally by the Navy as a conventional, but very advanced, turbojet engine providing extraordinary high thrust with the ability to operate routinely at high speeds up to Mach 3. The J-58 became increasingly difficult and expensive to develop for the A-12. Additionally, several of the aircraft for which the J-58 was originally intended had ceased to exist. As a result, the Navy lost interest and the J-58 was without a home or sponsor. On 30 January 1960 Lockheed received official word that funding for twelve A-12s had been approved. Pratt & Whitney also was informed they were cleared to move ahead with the construction of three "advanced, experimental engines for durability and reliability testing." Work could begin in earnest.

Work on the A-12's sensors also started. The primary camera manufacturer was Perkin-Elmer. Because of the extreme complexity of the design a decision was quickly made to fund Kodak's proposed backup system in case Perkin-Elmer ran into difficulty. At the same time, Minneapolis-Honeywell Corporation was selected to provide both the inertial navigation and automatic flight control systems. The David Clark Corporation became the prime sources of pilot equipment and associated life support hardware.

In February 1960 the CIA posed to Lockheed that it would screen a minimum of sixty Air Force pilots in an attempt to assemble an initial group of twenty-four A-12 pilots. Those selected would resign from the Air Force and be hired into the CIA as civilians, a process called "sheep dipping," similar to those who flew the CIA's U-2 aircraft. They also decided to put each pilot through a physical comparable to that of the Project Mercury astronauts.

In 1960, the USAF office of National Aeronautics and Space Administration (NASA) requested Ken Collins to volunteer for testing for a classified space program. He didn't know at the time, but he was volunteering for the highly classified A-12 Oxcart program. Ken tells how that came about:

The primary professional selection criterion of the Oxcart program was flying time in fighter type aircraft. The pilot needed more than a thousand operational flight hours in the aircraft and a total of two thousand flight hours. You had to be currently qualified and proficient. At that time I was an experienced instructor pilot in the RF-101, having had over fifty air-to-air refueling sorties and about ten transatlantic deployments, each with multiple air refuelings. That was the easy part for the Pentagon selection group. Either you had it or you didn't. The requirements to be highly qualified as an officer and a pilot was basic . . . the absolute minimum. All of your flight, professional and medical records were meticulously scrutinized at all levels before your name was released for further evaluation.

The personal requirement was that you had to be married and preferably that you had children. They were adamant about this after some problems they experienced with the previous U-2 program. Their explanation was that the family unit is more socially established, dedicated and dependable. Our wives were also interviewed separately, and psychological evaluations were conducted. Expanded background investigations were run on each wife as well.

At this phase (April 1961) of the overall evaluation, we still did not know what we were being evaluated for. The following events and schedules were generally the same for all the pilots being considered. Each was individually and separately tasked for the respective events. Initially I didn't know that there was another pilot (Captain Walter Ray) from Shaw Air Force Base (AFB) being evaluated. We started running into each other about a year later, when the field of pilots being considered was narrowed by process of elimination. From the beginning we were given the option of withdrawing from the selection process at any time without prejudice.

Our medical records were acceptable for the initial evaluation because we were all on flying status. However, Air Force physicals were not extensive enough for the final evaluation and we were about to find out what that really meant. I was scheduled for my "astronaut" physical at the Lovelace Clinic in Albuquerque, New Mexico. This is the same facility where the original astronauts received their medical evaluations, and was the medical facility for the original U-2 pilots. I discovered this later during a return visit to the clinic where I met Francis Gary Powers. The late Dr. William Lovelace, one of the pioneers of aviation medicine, founded the Lovelace Clinic.

I had a government contact (an Air Force flight surgeon) whom I met at the clinic. He established all of my schedules, appointments and observed all of the tests. He and I would have dinner together more as a part of the evaluation than just being social. The first medical phase lasted for five days. I arrived there on Sunday and departed the following Saturday. During those five days they checked out every bodily orifice, X-rayed every part of you from head to toe, flushed you out totally, took samples, and measured everything. I even carried a large brown bottle around for forty-eight hours to collect every drop of my urine. They conducted extensive EKGs and EEGs. I was hydrostatically weighted in a large water tank, ran the bicycle pulmonary functions, and passed

another physical stress test. I was then flown to the Los Alamos Laboratory in New Mexico to be inserted into the "body counter," which mapped the fat versus muscle tissue of my body. When correlated with the data from the hydrostatic weighting, theoretically, they could determine the total body capacity, regardless of size. After all that I was ready for the hospital!

From that date forward I was subjected to many different kinds of personal and professional evaluations. There was the soundproof black box where you had to remain for twelve hours in total darkness. Having various dinners and lunches with a whole variety of "professional" people. They were medical and intelligence people, ranging from senior managers to the basic company employee. All were there to get an opinion, except for the lesser guy. He was there to keep me out of trouble. After the sorting was near completion, I learned that eight would be selected for the initial program. This sounded reasonable since that was the average size group for the astronaut program, or so we thought!

I finally received my orders assigning me to the Headquarters, USAF, Washington, D.C., with a reporting date of 28 October 1962. I went back to Shaw AFB to start the moving process with my wife. While getting ready for the move, I was summoned to Washington, D.C., for a final meeting, at which time I still had the option of withdrawing from the program. For the first time, I was told it was not the astronaut program, but a project to fly and test an exotic new airplane for the CIA. There were no pictures or any other details.

I arrived in Washington, D.C., with my wife, Jane, and our four children. The next day I was taken to headquarters to sign on the dotted line. We headed west the following day. Jane had been interviewed and evaluated separately during the entire process. She was not told what I would be doing and was told not to talk about it to anyone. She could say that I decided to resign from the Air Force and go to work for Hughes Aircraft. For anyone who thinks we received a great salary, we all made about $4,000 more per year than our Air Force pay; however, the job was well worth it.

The first time I saw the A-12 was in December 1962 after I arrived at Area 51 [at Groom Lake in Nevada]. Colonel Doug Nelson, project manager, took me to a hangar and let me walk in by myself. What an amazing sight! The sun's rays entered the upper hangar windows, illuminating only the nose and spikes. As my eyes adjusted to the restricted light, I began to take in its sleek length, the massive twin rudders and its total blackness. A vision I will never forget.

Getting the family settled and feeling secure was important to the program. They knew the project pilots would generally be out of touch from Monday through Friday and some weekends. If there was a serious problem, my wife was given a telephone number to call any time and request that I call back. The return call could be immediate, or in one or two days, depending on my schedule and location. My wife knew that I was on a very special operational flying assignment, but not my location, what I was flying, my sponsor, or my associates, other than the project pilots. Once I had the family in place, I was ready to dedicate myself to the project.

When the final screening was completed the A-12 pilots included Ken Collins, Mele Vojvodich, Jack Layton, Dennis Sullivan, Francis Murray, Jack Weeks, William Skliar, Walter Ray, Alonzo Walter, David Young, and Russ Scott. Only the first six listed were destined to fly the A-12 on operational missions.

CHAPTER 2

A-12 Construction and Testing

In the summer of 1960, as the jigs began to go together inside the Burbank building where the aircraft were to be built, the difficulties entailed in the A-12's construction began to surface in ever-increasing numbers. The early decision to build the aircraft out of titanium to cope with the high temperatures at the Mach 3+ cruise speeds was a first for the aviation industry. Though numerous aircraft parts had been manufactured from titanium, an entire airframe had never been attempted. The A-12 was composed of 85 percent titanium and 15 percent composite materials. Though strong, relatively light, and capable of retaining its exceptional strength characteristics at very high temperatures, it also suffered from scarcity, high cost, and a reputation for being extremely difficult to work.

Other manufacturing challenges included the development of a fuel that could be safely stored in tanks where the temperature of the fuel reached 350 degrees F.; gaseous nitrogen inerting of the fuel tanks; a special lubricating oil that could be used effectively on parts that reached temperatures of 600 degrees F.; hydraulic seals capable of maintain their integrity at high temperatures; and a quartz glass window that could retain its optical qualities while being exposed to the extreme temperatures of the A-12's operating environment.

Another major problem was the radar-cross section (RCS). The areas causing the most RCS difficulty included the vertical stabilizers and the forward section of the engine nacelles. Lockheed conducted considerable research into the use of ferrites, high-temperature radar-absorbing materials and high-temperature plastic structures to find methods for reducing the RCS. Eventually, the vertical stabilizers, originally made of titanium, were replaced by high-temperature composites. This was almost certainly the first time such materials had been used for a major part of an aircraft's structure. Since the main domestic supplier, Titanium Metals Corporation of America, couldn't supply the entire project, titanium was sourced from all around the world—ironically, including the Soviet Union. Using third

Opposite: Groom Lake, also known as Area 51, was originally built to fly the CIA's U-2, but in later years took on a variety of top secret missions and aircraft testing. *Nevada Aerospace Hall of Fame Collection*

On 26 February 1962, the first A-12 made the trip from Burbank, California, to Groom Lake in specially built containers. The width of the trailers restricted the convoy's speed and required the cooperation of the highway patrol from California and Nevada. *Lockheed Martin*

parties and bogus companies, the CIA managed to purchase the necessary base metal from the Russians—who unwittingly helped the create the very plane that would be spying on them. It was the beginning of stealth technology.

The old U-2 location, Area 51, was prepared to accept the first A-12 aircraft. The considerable work needed to bring the facility up to standards was initiated in September 1960. In the interim, the runway was lengthened to 8,500 feet from the original 5,000 feet; the highway leading to the location was resurfaced; three surplus Navy hangars were erected at the north side; and an additional hundred surplus Navy buildings were moved to the base. Meanwhile, the prototype aircraft were being assembled with considerable difficulty at Burbank, California.

Excerpts from Kelly Johnson's log underscore some of the problems. March 6, 1961—"Having trouble with wing load distribution and have to put twist in outboard leading edge." During March, Johnson informed his CIA liaison that "schedules are in jeopardy on two fronts. One is the assembly of the wing and the other is in satisfactory development of the engine. Our evaluation shows that each of

intense at Burbank and work went on around the clock in three shifts. Engine problems continued as well. Following a meeting with Pratt & Whitney in Florida, Johnson notes in his log, "Their troubles are desperate. It is almost unbelievable that they could have gotten this far with the engine without uncovering basic problems which have been normal in every jet engine. Prospects of an early flight engine are dismal and I feel our program is greatly jeopardized. It's a good thing we went to the J-75, although these engines too have troubles and require new compressor discs."

During January of 1962 an agreement was reached with the Federal Aviation Administration (FAA) permitting the expansion of the restricted airspace surrounding the test location. Select FAA controllers were cleared for Oxcart operations. Additionally, nearby military radar facilities were briefed not to report radar sightings of high-performance aircraft.

The first A-12 was ready for final assembly during mid–February 1962. It was decided that, in light of the secrecy surrounding the project, the aircraft could not be flown out of Burbank. A special trailer was built to haul the aircraft across country. A thorough survey of the route during June 1961 assessed the hazards of moving the actual aircraft and showed that a package measuring 35 feet wide and 103 feet long could be moved along the road without major difficulty. Several road signs would have to be removed and selected trees would have to be trimmed. The entire fuselage, minus the wings, was crafted, covered with canvas, and loaded on the special trailer. The convoy left Burbank at 2:30 a.m. and arrived at the test location at 1:00 p.m.

After its arrival at Area 51, reassembly of the first aircraft and installation of the J-75 engines began. On 25 April 1962, the aircraft appeared ready for its initial test flight. Skunk Works test pilot Lou Schalk, who joined the program two years earlier, had spent many hours in the cockpit and a rudimentary simulator. By the advent of the

these programs is from three to four months behind the current schedule."

The CIA's Deputy Director for Plans Richard Bissell replied, "I have learned of your expected additional delay in first flight from 30 August to 1 December 1961. This news is extremely shocking on top of our previous slippage from May to August and my understanding as of our meeting 19 December that the titanium extrusion problems essentially overcome. I trust this is the last of such disappointments short of a severe earthquake in Burbank."

Unfortunately Johnson's problems continued. Construction pressures were

On 26 April 1962, Lockheed
chief test pilot, Lou Schalk,
takes to the skies with the first
flight in A-12, tail number 924, at
Groom Lake. The trouble-free
flight lasted thirty-five minutes.
Seven days later, on the third
test flight, Lou broke Mach 1.
Lockheed Martin

first A-12 taxi and previous flight testing, Lou was as prepared as possible. He recounts his first flight:

It had a very light load of fuel so it sort of accelerated really fast. Not many people really knew that we were going to lift the airplane off . . . some of the people in flight test engineering did know this. I was probably 3–4 percent behind the aft limit center of gravity (cg) when I lifted off the airplane . . . so it was unstable . . . and I fought it longitudinally which translated into lateral and directional problems, too.

The airplane wallowed through the sky until I finally felt I had it under control

enough to put it back down on the ground . . . which I did. By that time I was over the lakebed. This developed a big cloud of dust and the tower wanted to know if everything was OK. I said yes . . . but I have to roll out here on the lakebed and turn around and come back. They couldn't hear me because the Ultra High Frequency (UHF) transmitting antenna was on the bottom of the fuselage and that blanked out my transmission . . . so no one heard what I was saying. They asked me again and I said there was no problem. I wasn't trying to stomp on the brakes or anything. Everyone was having a heart attack. I finally made the turn and came out of the cloud of dust and they saw I hadn't run into the mountains on the other side of the lake and blow up the airplane. There was a huge sigh of relief!

I still didn't know what the devil was wrong with the airplane . . .why it handled so poorly . . . it didn't do that in the simulator and Kelly and I were talking about what we should do if we fly the airplane the next day. I said I think we should turn the dampers on . . . they were off when I was taxiing. On the second flight we were going to fly the plane around with the gear down.

On 26 April 1962 Lou Schalk flew the A-12 on its first scheduled airborne test flight. Kelly Johnson wrote, "We decided to fly with the stability augmentation system engaged on April 26. Everyone was awake just about through the night. We rolled out early and at 7:05 a.m. took off, making a beautiful takeoff. In flight we investigated the effects of the stability augmentation. We showed that the first flight troubles were not caused by basic aircraft stability."

Throughout 1962 more A-12 aircraft were trucked to Area 51. In November a two-seat trainer model of the A-12 arrived. It had an elevated second seat in the position behind the cockpit, normally occupied by sensor gear. Troubles still plagued the J-58 engine development. The trainer was equipped with J-75 engines so that flight training could be accommodated as expeditiously as possible.

In October 1962 the Cuban missile crisis reinvigorated the A-12 program. The loss of Maj. Rudolph Anderson's U-2 over Cuba on 27 October underscored the increasing vulnerability of this subsonic platform when operating in denied airspace. Successful execution of Oxcart now became a matter of the highest national priority. At the end of 1962, two A-12s were in flight test, one powered by J-57s, and the other with one J-57 and one J-58 engine. A speed of Mach 2.16 and an altitude of 60,000 feet had been achieved. By the end of January 1963, ten J-58s had been delivered to the

test location. The first A-12 flight with two J-58 engines took place on 13 January 1963. After that flight all the other A-12s were retrofitted with J-58s as well as the A-12s coming off the production line.

On 24 May 1963, the third A-12, involved in a subsonic engine test flight and piloted by Ken Collins, crashed fourteen miles south of Wendover, Utah. Ken successfully ejected and was unhurt. The A-12 fleet was temporarily grounded while an investigation was conducted. A pitot-static system failure due to icing was determined to be the culprit. In mid-October Kelly Johnson wrote, "We have been to Mach 3.0 twice, the first time being on 20 July 1963. On the second flight we blew an engine at design speed. It was very difficult to slow down and it rattled Lou Schalk around for three minutes. The aircraft stability augmentation system (SAS) did precisely as I asked it to do three years ago and no high structural loads were obtained."

Soon, another Lockheed test pilot, Jim Eastham, took the first A-12 out to a speed of Mach 3.3 and then cruised at Mach 3.2 for fifteen minutes. By the end of 1964 the A-12 fleet, consisting of eleven aircraft, had

Lou Schalk being congratulated after the perfect first flight test of the A-12. Right to left: Larry Bohanon, manager of A-12 flight testing at Groom Lake; CIA Deputy Director for Plans Rich Bissell; Schalk; Najeeb Halaby, administrator of the Federal Aviation Administration (FAA), and Joseph Charyk, director of the National Reconnaissance Office (NRO). *Lockheed Martin*

Habus attend the thirty-ninth anniversary of the first flight of the A-12 at the Virginia Aviation Museum on 28 April 2001. Lou Schalk (holding his coat) was on hand to talk about the first flight. *David Allison*

logged over 1,214 flights and 1,669 hours of flying time. Only six hours and twenty-three minutes had been logged above Mach 3.0 and only thirty-three minutes at Mach 3.2.

One flight was sent out to demonstrate complete systems reliability at Mach 3.05 and at 2,300 nautical miles (NM) range at an altitude of 76,000 feet. On 27 January 1965, an A-12 embarked on the first long-range, high-speed flight. Airborne for an hour and forty minutes, it spent an hour and fifteen minutes at Mach 3.1 or above. The mission covered 2,580 NM while cruising between 75,600 and 80,000 feet. During the tests that ensued, the A-12 achieved a maximum speed of Mach 3.29, an altitude of 90,000 feet, and a sustained flight time above Mach 3.2 of one hour and fourteen minutes. The longest flight lasted six hours and twenty minutes. During 1965, the "critical situation" the CIA's A-12 detachment had been waiting for finally emerged in Asia. On 18 March, the CIA's director, John A. McCone, discussed with Secretary of Defense Robert McNamara and Deputy Secretary Cyrus Vance the increasing hazards to U-2 and unmanned drone reconnaissance of Communist

China. It was further agreed that the necessary steps to begin the operation of flying Oxcart over China should proceed, with the construction and necessary arrangements on Okinawa, Japan. No decision was taken to fly the A-12s over China; this decision could only be made by the president.

The operation to fly A-12s out of Kadena Air Base, Okinawa, Japan, was codenamed Black Shield. The deployment of surface-to-air missiles around Hanoi also had caught U.S. forces off guard in the rapidly escalating Vietnam War. Secretary of Defense McNamara called this to the attention of the under secretary of the Air Force and inquired about the practicability of substituting A-12s for U-2s. He was told that Black Shield could operate over Vietnam as soon as adequate aircraft performance was achieved.

On 20 November 1965, Kelly Johnson wrote to the CIA, ". . . over all, my considered opinion is that the aircraft can be successfully deployed for the Black Shield mission with what I would consider to be at least as low a degree of risk as in the early U-2 deployment days. It is probably much less risky than our first U-2 deployment. I think the time has come when the bird should leave its nest." However, despite all its achievements, the A-12 was not called into operational duty.

Throughout 1966 the A-12 program in effect had nothing to do. Repeated requests to authorize the flights over China and North Vietnam were refused. The CIA, the Joint Chiefs of Staff, and the president's Foreign Intelligence Advisory Board favored the overflights while Alexis Johnson of the State Department, McNamara, and Vance opposed them. Perhaps most importantly, the president sided with the opposition, agreeing that the technological and political risks were too great.

On 12 May 1966 Kelly Johnson wrote, "As of this date, there is still no go-ahead for the deployment, although it seems fairly optimistic. The airplanes are ready to go." On 10 October 1966 he noted, "Still no deployment. We are making forty flights a month. The airplane is working quite well." An impressive demonstration of the A-12's capability occurred on 21 December 1966 when test pilot Bill Park flew the aircraft 10,198 statute miles (SM) in six hours. It was a record, unapproachable by any other aircraft in the world.

The CIA's inability to justify using the A-12 operationally by the end of 1966 had led to concerns the aircraft would soon be placed in storage, which accelerated during the first few months of 1967. On 21 March 1967, Kelly Johnson wrote, "The Air Force and the Program Office are having quite a time about storing the A-12s. It appears that half of them will be stored by the middle of this year and all of them by February of 1968. In the meantime, five airplanes will be kept on alert status for deployment."

With this bleak prognosis, it came as somewhat of a surprise during early May when prospects for operational deployment suddenly took a new turn. Concerns that SAMs would be deployed in North Vietnam suddenly provided the small A-12 community with a glimmer of hope. The president asked for a review of the surface-to-air activity and what might be done to monitor it. The CIA suggested using the A-12s. It was noted that the A-12 camera was far superior to those then being used on the various drones and the U-2s over North Vietnam.

As a result, State and Defense Department members reexamined the requirements and political risks of using the A-12s and the CIA's new director, Richard Helms, submitted another formal proposal to deploy the A-12s. President Johnson granted permission for use of the A-12 aircraft, and, on 16 May 1967, his assistant, Walt Rostow, formally conveyed the president's decision. Black Shield was put into effect immediately.

On 17 May 1967, the first support components of Black Shield were airlifted to Kadena Air Base (AB) in Okinawa. They were followed on 22 May by the first A-12,

which arrived six hours after departing the United States. A second A-12 followed on 24 May, and a third on 26 May. On 29 May, the unit was declared ready to fly its first operational mission. Under the command of Col. Hugh Slater, 260 personnel had been transferred to Kadena as part of the Black Shield team.

On 31 May the first A-12 operational sortie was flown. The flight followed a route over North Vietnam and the Demilitarized Zone (DMZ). It lasted three hours and thirty-nine minutes. Results were satisfactory and 70 of the 190 surface-to-air missile sites in North Vietnam were photographed, along with nine other priority targets. No radar signals were detected, implying that the Chinese and North Vietnamese had not known about the A-12 presence.

Over the next year, twenty-nine A-12 sorties were flown, twenty-six over North Vietnam, three over North Korea by Dennis Sullivan, Ken Collins, Francis "Frank" Murray, Mel Vojvodich, Jack Layton, and Jack Weeks. On 23 January 1968, the USS *Pueblo*, a Navy ship designed to intercept and sample the electronic environment off the east coast of North Korea, was seized in the middle of the night. Immediately, a CIA plan was implemented to fly the A-12s over North Korea. On 26 January,

The four A-12 hangars built by the CIA on Kadena AB, Okinawa, Japan. The elongated center structure housed all of the various maintenance shops. The same facilities were used for SR-71 operations. *USAF*

Above: All ten remaining A-12 aircraft parked at Groom Lake, Nevada, after the Oxcart program was terminated. Second from the left is the dual-seated A-12, nicknamed the "titanium goose," used to train A-12 pilots. It was never configured with J-58 engines. *Lockheed Martin*

Right: Kelly Johnson in the front seat of the "titanium goose," his only A-12 flight. Kelly has a standard Air Force helmet and flight suit on. Unfortunately, he never experienced a Mach 3+ flight because the plane was not capable of those speeds with two J-57 engines. *Lockheed Martin*

CIA pilot Jack Weeks' reconnaissance package found the *Pueblo*. This incident clearly demonstrated the ability of the A-12 to respond with minimal lead times to international incidents of political and military importance.

The last A-12 mission was on 8 May 1968, and the Kadena Detachment was then advised to prepare to deploy back to the United States. In May, Kelly Johnson wrote, "The decision was taken to phase out the A-12 by about mid-June. Plans were put into effect for storing the A-12 aircraft at Palmdale, CA." During March of 1968, SR-71s began arriving at Kadena to take over the Black Shield commitment. On 16 May, the Secretary of Defense, Clark Clifford, reaffirmed a previous decision to terminate and store the A-12 aircraft. On 21 May, President Johnson confirmed the decision. In gradual stages the A-12 was placed on standby to back up the SR-71. A date of 8 June was selected as the earliest possible date to begin redeployment of the A-12s back to the United States. By now the A-12 flights from Okinawa were limited to those essential for maintaining pilot proficiency.

During the final days overseas, the A-12 program suffered a final blow that was as inexplicable as it was tragic. On 4 June 1968, an A-12 piloted by Jack Weeks set out from Kadena on a check flight necessitated by an engine change. Weeks was last heard from when his plane was 520 miles east of Manila. Search and rescue operations found nothing. No cause for the accident was ever ascertained and his disappearance remains a mystery to this day.

The Memorial Wall at the entrance to CIA headquarters. Each of the 103 engraved stars on the marble wall represents a CIA officer killed in the line of duty. The Book of Honor lists seventy-nine names; the remaining twenty-four must remain secret, even in death, and are represented by a star in the book. Jack Weeks and Walt Ray are listed in the books. *CIA*

YF-12 Interceptor

Concurrent with the A-12's development, another version was being investigated. The YF-12A, referred to initially by Kelly Johnson as the AF-12, had come into being during the fall of 1960 as the end product of a proposal to the Air Force for a long-range, high-speed interceptor to counter newly perceived Soviet threats. The Soviets were developing and improving on their Bear and Bison nuclear bombers, giving them the capability to reach the United States. The Air Force was interested in developing a high-speed, high-altitude interceptor with a fire control system that had a look-down, shoot-down missile capability.

Until mid-1962, Johnson continued to work on both the A-12 and AF-12 programs simultaneously. This was an intensive and difficult undertaking, primarily because the two programs were mutually exclusive, and the security surrounding each required they be handled with great delicacy and little overlap. The seventh A-12 was marked to become the AF-12 prototype. The AF-12, as originally conceived by the Skunk Works, was a modified A-12 incorporating the fire control system integrated with a Hughes radar. A second seat in the original A-12 sensor system bay was added to accommodate a fire control system operator. In December of 1960, a separate project group was organized in the Skunk Works, working independently of the A-12 team. On January 23 and 24, the first meeting with the Air Force's weapon system project office took place at Burbank, briefing everyone on the Skunk Works' aircraft design and development philosophy.

By June, AF-12 wind tunnel tests had revealed directional stability problems that resulted from the revised nose and cockpit configuration mandated by the massive Hughes AN/AGS-18 radar. For the first time ventral fins were introduced into the design, two in the form of fixed surfaces attached to the underside of each engine nacelle and the third, a large folding fin mounted on the fuselage centerline. The AN/ASG-18 was the first U.S. coherent pulse Doppler radar design for long-range, look-down, or look-up detection of a Soviet aircraft. It consisted of forty-one units and weighed 1,380 pounds.

The YF-12 rear cockpit contained the radar controls (lower right) and missile controls (lower left), as well as the large radar scope and map displays. *Lockheed Martin*

Opposite top: The YF-12 painted with three speed records set on May 1, 1965. The 800-pound Hughes AIM-47 was a Mach 4 radar-guided air-to-air missile with a range in excess of 125 miles. Notice sideways ventral fin at the rear of the YF-12; once the landing gear retracted on takeoff, the ventral fin folded downward to provide directional stability. *Lockheed Martin*

Opposite bottom: Loading the Hughes AIM-47 into one of the three bays on the YF-12. Its technology was later applied to the highly successful AIM-54 Phoenix missile used in the Navy's F-14 Tomcats. *Lockheed Martin*

The AF-12's fire control system and missile complement was a major concern for Johnson. No one had ever fired an air-to-air missile while flying at Mach 3 speeds, and there seemed to be little agreement as to how to eject the missile from the AF-12's weapon bay. Even the design of the missile pylon and trapeze assembly was ill-defined, and the company overseeing the fire control system, Hughes Aircraft, was as much in the dark as Lockheed. Paper studies verified it could be done, and wind tunnel tests confirmed these, but it would only be through testing of actual hardware that realistic insight could be gained.

From December of 1962 through the early spring, construction of the three AF-12s in a segregated corner of the Burbank manufacturing facility moved ahead smoothly. It was decided that the initial flight testing would take place at the A-12 test location. On 7 August 1963, the first AF-12, with Lockheed's test pilot Jim Eastham at the controls, made its maiden flight. Flight testing proceeded without significant difficulty, and during January 1964 the aircraft was temporarily grounded while intake system upgrades and newer engines with increased thrust were installed.

On 29 February 1964, President Johnson announced,

The United States has successfully developed an advanced experimental jet aircraft, the A-11, which has been tested in sustained flight at more than two thousand miles per hour and at altitudes in excess of 70,000 feet.

The performance of the A-11 far exceeds that of any other aircraft in the world today. The development of this aircraft has been made possible by major advances in aircraft technology of great significance to both military and commercial applications.

Several A-11 aircraft are now being flight tested at Edwards AFB. . . . In view of the continuing importance of these developments to our national security, the detailed performance of the A-11 will remain strictly classified and all individuals associated with the program have been directed to refrain from making any further disclosure concerning the program.

His reference to the A-11 was of course the result of Kelly Johnson's input. There was no A-11, at least in hardware form, and the photographs released in conjunction with the president's announcement actually depicted the first AF-12. Compounding the confusion was the fact the AF-12 now was assigned an official Air Force designation of YF-12A. Flight testing continued with increased frequency, as confidence in the aircraft grew. On 16 April 1964, the first AIM-47 air-to-air missile was launched in flight. According to Kelly Johnson, "The launching was safe but the angle developed was poor. If it had been a powered missile, it would have come out through the cockpit." On 28 September 1964, an AIM-47 missile was fired from a YF-12A at Mach 3.2 at 75,000 feet. Flying at 40,000 feet and at a range of thirty-six miles, it missed its target by six feet, which was considered within the kill zone.

During mid-1964, some difficulties with the A-12, also reflected in the YF-12, had yet to be fully overcome. Transonic

acceleration remained troublesome and miscellaneous sub-systems still required major improvements in dependability. On 9 July 1964, test pilot Bill Park was forced to eject from an A-12 during final approach to landing at the test location. Bill was uninjured and the cause of the accident was quickly determined to be a stuck outboard elevon servo valve.

Between March 1965 and September 1966, actual flight testing of the fire control system was conducted at Eglin AFB, Florida, against real targets in the Gulf of Mexico. Seven missile firings took place, only one of which was deemed unsuccessful. During July 1966, Kelly Johnson wrote, "We were directed to give up further flying of the FY-12s, though we had proposed shooting down a drone at Holloman to get the effect of ground clutter for low altitude targets." On 5 January 1968, Kelly Johnson received an official notice to close down the YF-12 program by 1 February 1968.

CHAPTER 4

D-21 Drone Project

Opposite: The D-21 drone has very little clearance between the two rudders of the M-21. During one launch, the drone came in contact with the M-21, causing Lockheed test pilot Bill Park and launch control engineer Ray Torick to eject over the Pacific Ocean. Park was rescued, but tragically, Ray Torick drowned in the Pacific Ocean. *Lockheed Martin*

Below: A full-scale mock-up of the D-21 drone. Black areas were fabricated from composite materials to lower the drone's radar cross section. The actual drone differed only in detail from this mock-up. *Lockheed Martin*

Of the various Skunk Works Mach 3 aircraft programs, the least known to reach the operational hardware stage was undoubtedly the D-21 unmanned strategic reconnaissance drone. Developed and operated by the CIA and Air Force under extreme secrecy, its existence was unveiled early in 1977. The D-21 program was an extension of the A-12 program, brought to life in response to the U.S. government's decision to discontinue manned overflights after the loss of Gary Powers and his U-2 on 1 May 1960. On 10 October 1962, the Skunk Works received authorization for a drone study from the CIA.

A full-scale mock-up was completed on 7 December 1962. It included the "anti-radar" characteristics of the actual drone. Propulsion system work began to accelerate at this time. A Marquardt RJ43-MA-3 Bomarc engine was wind-tunnel tested at simulated D-21 operating conditions. A minor dimension change in the drone forced Hycon to redesign their cameras. On 6 August 1963 the change was deemed successful and had not impacted the quality of the imagery.

By October, the drone's overall configuration had been finalized and the equally difficult task of defining the A-12 launch system was nearing an end. The A-12 launch aircraft was now renamed the M-21, "M" standing for "Mother" ship. The drone became the D-21, the "Daughter" aircraft. The numerals were simply reversed so as not to confuse the "mother/daughter" combination.

Concurrently with the D-21 work, building of the two M-21s to serve as launch platforms was also underway. A single, dorsally mounted pylon had been developed that was low drag, yet sufficiently strong enough to

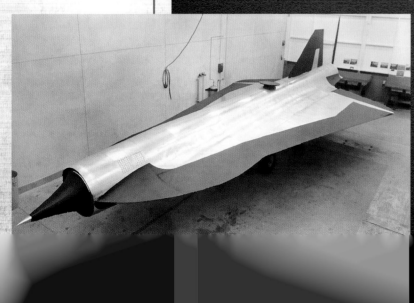

Test mounting of the D-21 on the dorsal centerline pylon of an M-21 was completed on 19 June 1964. Safety concerns of transporting the drone up to Mach 3 were significant. The supporting pylon had to be shortened for structural considerations, necessitating a pushover maneuver for the launch. *Lockheed Martin*

support the eleven thousand-pound D-21 at Mach 3. The pylon contained a series of latches that secured the D-21 to the M-21, provisions for emergency pneumatic jettison of the D-21, and a refuel line that was used to top off the D-21's tanks prior to launch. D-21 separation was to be accomplished by flying the drone off the M-21 during a slight (approximately 0.9 g) pushover maneuver.

A fit check of the D-21 drone mated to the M-21 aircraft took place on 19 June 1964. A few problems surfaced, but on 12 August 1964 the drone was sent to the test location to undergo initial flight testing. On 22 December, the first flight of the M-21/D-21 combination was successfully completed. (Interestingly, on this same day, the first SR-71 successfully completed its first flight from Lockheed's Palmdale, California, facility.) However, there

were still major problems with the star tracker, used for navigation of the D-21 once launched. By May, the two-aircraft combination had been flown out to Mach 2.6, but without a launch.

On 21 October 1965, Kelly noted, "We have had great difficulties in getting the M-21 to speed and range. We tried to run to Point Mugu for launch practice, but couldn't make the range. Transonic acceleration is very poor, particularly with hot temperatures. As of this day, we are putting in 34,000 pound thrust engines. We are driving to a launch date of 15 November." Further problems delayed the first launch date to 5 March 1966. According to Johnson, it was a great success. They demonstrated the most dangerous maneuver, the actual launch, which worked perfectly. The D-21 flew 120 miles and was lost in the Pacific Ocean.

On 27 April 1966, the second launch was equally as successful. Kelly writes, "I went to Pt. Mugu for the second launching . . . this was a dandy flight, going over 1,200 NM and holding course within half mile for the whole flight. It reached 90,000 feet, Mach 3.3, and finally fell out of the sky when a hydraulic pump burned out. At this time I have proposed the use of the D-21 with a rocket launch from the B-52. Our problem now will be to get wide usage of this new bird at low cost." A second batch of fifteen D-21s was ordered on 29 April 1966, and the following month Johnson made a formal proposal to Strategic Air Command (SAC) to launch the drone from Boeing B-52Hs. This was based on greater safety, lower cost, and greater deployment range.

On 16 June 1966 a third D-21 drone was successfully launched. It flew about 1,600 NM and made eight programmed turns, but did not successfully eject the photographic package due to an electronic failure. The

fourth flight on 30 July was a total disaster. The M-21, piloted by Bill Park and with Ray Torick in the launch control operator's position, was lost over the Pacific Ocean when the D-21 collided with the M-21. Observed by another M-21, which was flying chase, the D-21 and M-21 were seen to make contact shortly after release. At Mach 3.25, the M-21 pitched up and its nose broke off. Park and Torick ejected, but by the time a rescue helicopter arrived on the scene, Torick had drowned because his pressure suit filled with water. Bill Park was recovered 150 miles at sea after floating for an hour.

Because of the safety issue and losing one of his test engineers, Kelly Johnson cancelled the M-21/D-21 program and pursued launching the drones from the B-52H aircraft. In its proposed operational configuration, the D-21 was to be launched from a B-52H after arriving at any launch destination. The capabilities of the system were made possible by the

Part of the M-21 mission profile included inflight refueling to extend range of the D-21. The 11,000-pound D-21 was not only heavy, but also adversely affected the M-21's normal performance due to its high drag. Note that the D-21's nose inlet has a cone covering. *Lockheed Martin*

Above: One of Beale's two B-52H models modified to carry the D-21 loaded with both drones during the testing phase. The long-range capability of the B-52H allowed launching of the D-21 drone from anywhere in the world, well outside of enemy territory. *Lockheed Martin*

Right: The D-21's delta wing was severely damaged during the first, and last, attempt to pyrotechnically remove the aerodynamic cone faring from the D-21's intake. By discarding the nose cone, the D-21's ramjet engine could be used to supplement the M-21's J-58 engines. *Lockheed Martin*

The Director
cordially invites you
to a ceremony and reception
commemorating the presentation of
an A-12 OXCART Reconnaissance Aircraft
on the occasion of the
60th Anniversary of the Central Intelligence Agency
on Wednesday, the nineteenth of September
at two o'clock in the afternoon
Central Intelligence Agency
George Bush Center for Intelligence
McLean, Virginia

Military-Uniform of the Day

R.S.V.P. 703.482.7291
Please provide Social Security Number
Please arrive by 1:30 p.m. via the Route 123 Entrance Gate
Photo Identification Required

3,000 NM range of the D-21, and the B-52H's ability to fly to virtually any spot on the globe.

Following an air launch, the D-21 was accelerated to a Mach 3.3+ and an altitude of 80,000 feet by a solid propellant booster rocket developed by Lockheed Propulsion Company. The booster was then jettisoned and the D-21 commenced on a 3,000 NM cruise over a preprogrammed course. At the end of the flight, a controlled descent was made to a lower altitude where the hatch payload was ejected and lowered by parachute to an altitude where an air retrieval was executed by a C-130B

Hercules aircraft. The recoverable payload contained the reconnaissance camera, the inertial navigation system, the automatic flight control system, the command and telemetry electronics, the recovery beacons, and the recovery parachute system.

On 9 November 1969, the first operational mission from a B-52H was launched but did not succeed. The D-21 simply disappeared. The second and third operational launches were of little value in providing intelligence. On 20 March 1971, the fourth and final operational mission was flown and never recovered. This ended the D-21 drone program.

Left: The CIA invitation to unveiling of A-12 tail number 931 at their headquarters. It was originally on display at the Minnesota Air National Guard Museum in St. Paul and shipped to CIA headquarters in 2007. *Rich Graham*

Top: Former CIA Director Gen. Michael V. Hayden (center) and Mrs. Hayden speak with Oxcart pilot Dennis Sullivan at the A-12's dedication at CIA headquarters. *CIA*

Bottom: Lockheed test pilots (left to right) Lou Schalk (A-12), Jim Eastham (YF-12), and Bob Gilliland (SR-71) each flew a Blackbird on its first flight. In April 2002, they celebrated the fortieth anniversary of the A-12's first flight. *Tony Landis*

CHAPTER 5

Enter the SR-71

Opposite: SR-71 production facility at Burbank. Aircraft in foreground is tail number 980, which had its first flight on 25 September 1967. *Lockheed Martin*

Below: The rounded features make it highly improbable for a ground-tracking radar to track and lock onto the SR-71 at 80,000 feet. *Lockheed Martin*

During the course of the A-12 program, the Air Force had been exceedingly helpful to the CIA. It provided financial support, conducted the refueling program, provided operational facilities at Kadena, and airlifted A-12 personnel and supplies to Okinawa for the operations over Vietnam and North Korea. Through it all, the Air Force remained frustrated that a strategic reconnaissance mission had been given to another government agency. During March of 1962, the Air Force had been cleared to give Lockheed a study contract, the basis of which would be the CIA's A-12 aircraft.

Several aircraft designs were developed by the Skunk Works during this time. A reconnaissance/strike variant of the A-12, the RS-12, was one of them. Another version of the A-12, the R-12, designed to be a utilitarian/universal aircraft, appealed to Kelly Johnson. With the advent of the R-12, the CIA had begun to diverge into other sensor platforms of a considerably more clandestine nature, including the first of the super-high resolution optical systems being merged with satellite technology. As the A-12s had been supported by Air Force expertise, it was not surprising when the CIA capitulated to Air Force demands, returning the airborne strategic reconnaissance mission to their jurisdiction.

The R-12 would differ from the A-12 in several ways. It had a pressurized rear cockpit for a second Air Force crew member, the fuselage was stretched slightly to accommodate more fuel, and the sensor's locations were changed to be on both side of the forward fuselage.

On 24 July 1964, President Johnson made the first official announcement concerning the forthcoming Lockheed SR-71. As part of the A-12 history, it is necessary to understand that from this point forward, there was a plan to replace the CIA's A-12 with the Air Force's SR-71.

On 29 October 1964, the first SR-71 was delivered by truck from Burbank to Palmdale for final assembly and preflight preparations. On 22 December 1964, Skunk Works test pilot Bob Gilliland took the first SR-71 on its maiden flight.

ON JULY 24, 1964, President Lyndon B. Johnson gave a news conference at the State Department auditorium with 315 people in attendance. For years I had heard that President Johnson became dyslexic when he announced for the first time that the United States had developed a Mach 3 reconnaissance aircraft. Legend says that President Johnson was supposed to name the Blackbird the "RS-71" and that he confused the letters, reversed them and said "SR-71."

In 2000 I decided to research this. I contacted the Lyndon Baines Johnson Library in Austin, Texas. Mr. Michael Parrish, the archivist there, sent me a copy of the speech LBJ made that day in 1964:

Good afternoon, ladies and gentlemen. I would like to announce the successful development of a major new strategic manned aircraft system, which will be employed by the Strategic Air Command. This system employs the new SR-71 aircraft, and provides a long-range, advanced strategic reconnaissance plane for military use, capable of worldwide reconnaissance for military operations.

The Joint Chiefs of Staff, when reviewing the RS-70, emphasized the importance of the strategic reconnaissance mission. The SR-71 aircraft reconnaissance system is the most advanced in the world. The aircraft will fly at more than three times the speed of sound. It will operate at altitudes in excess of 80,000 feet. It will use the most advanced observation equipment of all kinds in the world. The aircraft will provide the strategic forces of the United States with an outstanding long-range reconnaissance capability. The system will be used during periods of military hostilities and in other situations in which the United States military forces may be confronting foreign military forces.

The SR-71 uses the same J-58 engine as the experimental interceptor previously announced, but it is substantially heavier and it has a longer range. The considerably heavier gross weight permits it to accommodate the multiple reconnaissance sensors needed by the Strategic Air Command to accomplish their strategic reconnaissance mission in a military environment.

This billion dollar program was initiated in February of 1963. The first operational aircraft will begin-flight testing in early 1965. Deployment of production units to the Strategic Air Command will begin shortly thereafter.

Appropriate Members of Congress have been kept fully informed on the nature of and the progress in this aircraft program. Further information on this major advanced aircraft system will be released from time to time at the appropriate military secret classification levels.

The remainder of his speech went on to other subjects.

In all three places where President Johnson's speech text stated "SR-71," the audio tapes reveal he said, "SR-71." The LBJ library also sent me a copy of the "Stenotype Transcript of Press Conference." In those days the White House used a stenographer writing in shorthand what he heard, or thought he heard, the president saying.

However, it was the stenographer who got dyslexic and in his transcript wrote "RS-71" in all three places where President Johnson said, "SR-71." The stenographer's version was handed out to the press reporters that day. I am sure the reporters looked at the handout and must have said, "I'm positive Johnson said SR-71 . . . this reads RS-71? He must have got it backwards." In actuality, it was the stenographer that got it backwards. However, in slight defense of the stenographer, President Johnson's first reference to the RS-70, a reconnaissance version of the ill-fated Valkyrie, most likely caused the confusion.

Refueling in conditions like this allowed crews to watch the sunset in the west, then see it rise again in the west as they began their acceleration and cruise leg and "caught up" with the setting sun.
Lockheed Martin

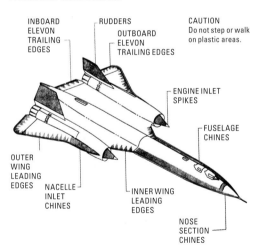

INBOARD ELEVON TRAILING EDGES

RUDDERS

OUTBOARD ELEVON TRAILING EDGES

CAUTION
Do not step or walk on plastic areas.

ENGINE INLET SPIKES

FUSELAGE CHINES

OUTER WING LEADING EDGES

NACELLE INLET CHINES

INNER WING LEADING EDGES

NOSE SECTION CHINES

Departing from Lockheed's Air Force Plant 42 facility at Palmdale, it remained airborne for just over an hour and reached a speed in excess of 1,000 miles per hour (mph). Ongoing flight testing with the SR-71 revealed problems with the fuel, hydraulic, and electrical systems, requiring extensive reworking. These problems were soon resolved.

One week before the first SR-71 flight, the Commander in Chief SAC (CINCSAC) Gen. John Ryan revealed that the 4200th Strategic Reconnaissance Wing at Beale AFB, California, would be the new home for the SR-71s. An $8.4-million construction program was quickly initiated at Beale.

On 25 January 1966, Lockheed test flight employees Bill Weaver and Jim Zwayer took off from Edwards AFB in SR-71 tail number 952. Although not a true "ejection" out of the SR-71, the following story told by Weaver is priceless:

Among professional aviators, there's a well-worn saying: Flying is simply hours of boredom punctuated by moments of stark terror. But I don't recall too many periods of boredom during my thirty-year career with Lockheed . . . most of which was spent as a test pilot.

By far, the most memorable flight occurred on 25 January 1966. Jim Zwayer, a Lockheed flight-test specialist, and I were evaluating systems on an SR-71 Blackbird test from Edwards. We also

Above: Preflight preparations prior to the first SR-71 flight on 22 December 1964. The massive Buick start cart is underneath the right engine ready for engine rotation. *Lockheed Martin*

Left: The plastic surfaces on the SR-71 (as indicated in the maintenance handbook) are not externally visible and could be easily damaged. *USAF*

Right: Lockheed's chief project pilot for the SR-71, Robert "Bob" Gilliland. On the first flight Bob achieved Mach 1.5 at 50,000 feet, a notable achievement for any aircraft's first flight. *Lockheed Martin*

F-104 flying chase during a low-altitude pass over Palmdale's runway on the first SR-71 flight by test pilot Bob Gilliland. *Lockheed Martin*

during supersonic flight to decelerate airflow in the duct, slowing it to subsonic speed before reaching the engine's face. This was accomplished by the inlet's center-body spike translating aft, and modulating the inlet's forward bypass doors.

Normally, these actions were scheduled automatically as a function of Mach number, positioning the normal shock wave (where airflow becomes subsonic) inside the inlet to ensure optimum engine performance. Without proper scheduling, disturbances inside the inlet could result in the shock wave being expelled forward—a phenomenon known as an "inlet unstart" (see chapter 16).

That causes an instantaneous loss of engine thrust, explosive banging noises and violent yawing of the aircraft—like being in a train wreck. Unstarts were not uncommon at that time in the SR-71's development, but a properly functioning system would recapture the shock wave and restore normal operation.

On the planned test profile, we entered a programmed thirty-five-degree bank turn to the right. An immediate unstart occurred on the right engine, forcing the aircraft to roll further right and start to pitch up. I jammed the control stick as far left and forward as it would go.

No response. I instantly knew we were in for a wild ride.

I attempted to tell Jim what was happening and to stay with the airplane until we reached a lower speed and altitude. I don't think the chances of surviving an ejection at Mach 3.18 and 78,800 ft. were very good. However, g-forces built up so rapidly that my words came out garbled and unintelligible, as confirmed later by the cockpit voice recorder.

The cumulative effects of system malfunctions, reduced longitudinal stability, increased angle-of-attack in the turn, supersonic speed, high altitude and other factors imposed forces on the airframe that exceeded flight control authority and the Stability Augmentation System's ability to restore control.

were investigating procedures designed to reduce trim drag and improve high-Mach cruise performance. The latter involved flying with the center of gravity (CG) located further aft than normal, reducing the Blackbird's longitudinal stability.

We took off from Edwards at 11:20 a.m. and completed the mission's first leg without incident. After refueling from a KC-135 tanker, we turned eastbound, accelerated to Mach 3.2 cruise speed and climbed to 78,000 feet, our initial cruise-climb altitude.

Several minutes into the cruise, the right engine inlet's automatic control system malfunctioned, requiring a switch to manual control. The SR-71's inlet configuration was automatically adjusted

The 4200th Strategic Reconnaissance Wing on the tarmac with the B-52s at Beale. *USAF*

Everything seemed to unfold in slow motion. I learned later the time from event onset to catastrophic departure from controlled flight was only two to three seconds. Still trying to communicate with Jim, I blacked out, succumbing to extremely high g-forces.

Then the SR-71 . . . literally . . . disintegrated around us.

From that point, I was just along for the ride. And my next recollection was a hazy thought that I was having a bad dream. Maybe I'll wake up and get out of this mess, I mused. Gradually regaining consciousness, I realized this was no dream; it had really happened. That also was disturbing, because . . . I could not have survived what had just happened.

I must be dead. Since I didn't feel bad, just a detached sense of euphoria, I decided being dead wasn't so bad after all. As full awareness took hold, I realized I was not dead. But somehow I had separated from the airplane.

I had no idea how this could have happened; I hadn't initiated an ejection. The sound of rushing air and what sounded like straps flapping in the wind confirmed I was falling, but I couldn't see anything. My pressure suit's face-plate had frozen over and I was staring at a layer of ice.

The pressure suit was inflated, so I knew an emergency oxygen cylinder in the seat kit attached to my parachute harness was functioning. It not only supplied breathing oxygen, but also pressurized the suit, preventing my blood from boiling at extremely high altitudes. I didn't appreciate it at the time, but the suit's pressurization had also provided physical protection from intense buffeting and g-forces. That inflated suit had become my own escape capsule.

My next concern was about stability and tumbling. Air density at high altitude is insufficient to resist a body's tumbling motions, and centrifugal forces high enough to cause physical injury could develop quickly. For that reason, the SR-71's parachute system was designed to automatically deploy a small-diameter stabilizing chute shortly after ejection and seat separation. Since I had not intentionally activated the ejection sequence, it occurred to me the stabilizing chute may not have deployed.

The protection provided by his silver pressure suit saved Bill Weaver's life when he was "ejected" from a disintegrating SR-71. *Lockheed Martin*

However, I quickly determined I was falling vertically and not tumbling. The little chute must have deployed and was doing its job. Next concern: the main parachute, which was designed to open automatically at 15,000 feet. Again I had no assurance the automatic-opening function would work.

I couldn't ascertain my altitude because I still couldn't see through the iced-up faceplate. There was no way to know how long I had been blacked-out or how far I had fallen. I felt for the manual-activation D-ring on my chute harness, but with the suit inflated and my hands numbed by cold, I couldn't locate it. I decided I'd better open the faceplate, try to estimate my height above the ground, then locate that D-ring.

Just as I reached for the faceplate, I felt the reassuring sudden deceleration of main-chute deployment.

I raised the frozen faceplate and discovered its up-latch was broken. Using one hand to hold that plate up, I saw I was descending through a clear, winter sky with unlimited visibility. I was greatly relieved to see Jim's parachute coming down about a quarter of a mile away. I didn't think either of us could have survived the aircraft's breakup, so seeing Jim had also escaped lifted my spirits incredibly.

I could also see burning wreckage on the ground a few miles from where we would land. The terrain didn't look at all inviting—a desolate, high plateau dotted with patches of snow and no signs of habitation.

I tried to rotate the parachute and look in other directions. But with one hand devoted to keeping the faceplate up and both hands numb from high altitude, subfreezing temperatures, I couldn't manipulate the risers enough to turn. Before the breakup, we'd started a turn in the New Mexico-Colorado-Oklahoma-Texas border region. The SR-71 had a turning radius of about a hundred miles at that speed and altitude, so I wasn't even sure what state we were going to land in. But, because it was about 3:00 p.m., I was certain we would be spending the night out here.

At about 300 feet above the ground, I yanked the seat kit's release handle and made sure it was still tied to me by a long lanyard. Releasing the heavy kit ensured I wouldn't land with it attached to my derriere, which could break a leg or cause other injuries. I then tried to recall what survival items were in that kit, as well as techniques I had been taught in survival school.

Looking down, I was startled to see a fairly large animal—perhaps an antelope—directly under me. Evidently, it was just as startled as I was because it took off in a cloud of dust.

My first-ever parachute landing was pretty smooth. I landed on fairly soft ground, managing to avoid rocks, cacti, and antelopes. My chute was still billowing in the wind, though. I struggled to collapse it with one hand, holding the still-frozen faceplate up with the other.

"Can I help you?" a voice said.

Was I hearing things? I must be hallucinating. Then I looked up and saw a guy walking toward me, wearing a cowboy hat. A helicopter was idling a short distance behind him. If I had been at Edwards and told the search-and-rescue unit that I was going to bail out over the Rogers Dry Lake at a particular time of day, a crew couldn't have gotten to me as fast as that cowboy-pilot did.

The gentleman was Albert Mitchell, Jr., owner of a huge cattle ranch in northeastern New Mexico. I had landed about a mile and a half from his ranch house—and from a hanger for his two-place Hughes helicopter. Amazed to see him, I replied I was having a little trouble with my chute. He walked over and collapsed the canopy, anchoring it with several rocks. He had seen Jim and me floating down, and had radioed the New Mexico Highway Patrol, the Air Force, and the nearest hospital.

Extracting myself from the parachute harness, I discovered the source of those flapping-strap noises heard on the way down. My seat belt and shoulder harness were still draped around me, attached and latched. The lap belt had been shredded on each side of my hips, where the straps had fed through knurled adjustment rollers. The shoulder harness had shredded in a similar manner across my back. The ejection seat had never left the airplane; I had been ripped out of it by the extreme forces, seat belt and shoulder harness still fastened.

I also noted that one of the two lines that supplied oxygen to my pressure suit had come loose, and the other was barely hanging on. If that second line had become detached at high altitude, the deflated pressure suit wouldn't have provided any protection. I knew an oxygen supply was critical for breathing and suit-pressurization, but didn't appreciate how much physical protection an inflated pressure suit could provide.

That the suit could withstand forces sufficient to disintegrate an airplane and shred heavy nylon seat belts, yet leave me with only a few bruises and minor whiplash, was impressive. I truly appreciated having my own little escape capsule.

After helping me with the chute, Mitchell said he'd check on Jim. He climbed into his helicopter, flew a short distance away and returned about ten minutes later with devastating news. Jim was dead. Apparently, he had suffered a broken neck during the aircraft's disintegration and was killed instantly.

Mitchell said his ranch foreman would soon arrive to watch over Jim's body until the authorities arrived. I asked to see Jim and, after verifying there was nothing more that could be done, agreed to let Mitchell fly me to

Pilot Jim Watkins (left) and his RSO David Dempster at the completion of their fifth operational sortie in 1968. Although not shooting down enemy planes, five enemy "kills" and some literary license allowed them to be "SR-71 Aces." *Bill Gornik*

the Tucumcari hospital, about sixty miles to the south.

I have vivid memories of that helicopter flight, as well. I didn't know much about rotorcraft, but I knew a lot about "red lines," and Mitchell kept the airspeed at or above red line all the way. The little helicopter vibrated and shook a lot more that I thought it should have. I tried to reassure the cowboy-pilot I was feeling OK; there was no need to rush. But since he'd notified the hospital staff that we were inbound, he insisted we get there as soon as possible. I wouldn't help but think how ironic it would be to have survived one disaster only to be done in by the helicopter that had come to my rescue.

However, we made it to the hospital safely—and quickly. Soon, I was able to contact Lockheed's flight test office at Edwards. The test team there had been notified initially about the loss of radio and radar contact, then told the aircraft had been lost. They also knew what our flight conditions had been at the time, and assumed no one could have survived. I explained what had happened, describing in fairly accurate detail the flight conditions prior to breakup.

The next day, our flight profile was duplicated on the SR-71 flight simulator at Beale AFB. The outcome was identical. Steps were immediately taken to prevent a recurrence of our accident. Testing at a CG aft of normal limits was discontinued, and trim-drag issues were subsequently resolved via aerodynamic means. The inlet control system was continuously improved, and, with subsequent development of the Digital Automatic Flight and Inlet Control System, inlet unstarts became rare.

Investigation of our accident revealed that the nose section of the aircraft had broken off aft of the rear cockpit and crashed about ten miles from the main wreckage. Parts were scattered over an area approximately fifteen miles long and ten miles wide. Extremely high air loads and g-forces, both positive and negative, had literally ripped Jim and me from the airplane. Unbelievably good luck is the only explanation for my escaping relatively unscathed from that disintegrating aircraft.

Two weeks after the accident, I was back in an SR-71, flying the first sortie on a brand-new bird at Lockheed's Palmdale, California, assembly and test facility. It was my first flight since the accident, so a flight test engineer in the back seat was probably a little apprehensive about my state of mind and confidence.

As we roared down the runway and lifted off, I heard an anxious voice over the intercom. "Bill! Bill! Are you there?" "Yeah, George. What's the matter?" "Thank God! I thought you might have left." The rear cockpit of the SR-71 has no forward visibility—only a small window on each side—and George couldn't see me. A big red light on the master-warning panel in the rear seat had illuminated just as we rotated, stating: "Pilot Ejected." Fortunately, the cause was a misadjusted micro switch, not my departure.

In addition to the steps taken to avoid similar aircraft failure in the future, Bill's inability to see out of his faceplate from ice buildup led to improved SR-71 ejection seats, which incorporated a battery pack that kept the glass faceplate heated during the frigid descent.

BOYLE'S LAW OF PHYSICS states that as outside pressure decreases around a sealed container, its volume has to increase. A balloon that floats higher in the sky becomes larger and larger as it gains altitude. When you eject from an SR-71 at high altitude, the pressure suit's sensitive pressure controller (located on the right side of the suit) closes up instantly, trapping the cockpit pressure of 26,000 feet inside the suit. Outside the aircraft at 80,000 feet, Boyle's Law is in effect and the pressure suit immediately inflates because of the inside/outside pressure differential. This is the rapid decompression (RD) all SR-71 crew members experienced and routinely practiced in the altitude chamber at Beale. Tom Bowen at the Beale PSD facility, who is considered "Mr. Pressure Suit," confirmed this is precisely what happens. He further added, however, after the initial inflation of the suit, it will continue to stay inflated on the way down by the crew member's normal breathing until a safe altitude is reached.

CHAPTER 6

Home for the SR-71 Blackbird: Beale AFB California

Beale AFB in northern California was selected to be home for the SR-71 aircraft. Its location about twenty-five miles north of Sacramento, California, isolated from major populated areas, made it an ideal place to fly the highly classified SR-71 from. There were over twenty-two thousand acres of land on Beale, and the long runway and excellent weather made it an ideal location for flying operations.

The 4200th Strategic Reconnaissance Wing (SRW) was activated at Beale AFB on 1 January 1965 in anticipation of the SR-71s' arrival. A year later, on 7 January 1966, wing commander, Col. Doug Nelson, along with instructor pilot Lt. Col. Ray Haupt, delivered the first SR-71 to Beale. It was the dual-controlled SR-71 trainer model, called the SR-71B model (tail number 956).

Beale AFB from 70,000 feet, taken by an SR-71. About a mile below the runway and flight line area is the cantonment area, and base housing is at the bottom left corner. The old Yuba Goldfields can be seen at upper right corner. *USAF*

Left: SR-71 on downwind for landing on runway 15 at Beale AFB. On the ramp area you can see twelve individual hangars for housing each SR-71. *Lockheed Martin*

Below: The SR-71B model trainer on the ramp at Beale. *Lockheed Martin*

After his Mach 3 flight in the SR-71, Gen. Chuck Yeager (first to break the sound barrier on 14 October 1947) is congratulated by the 9th Wing commander, Hector Freeze, at Beale AFB. *USAF*

The second B-model trainer followed shortly and "A" models continued to arrive steadily through late 1967.

The first two of eight T-38 Talon trainer aircraft arrived at Beale on 7 July 1965. The T-38 was to be used as a companion trainer to maintain overall flying proficiency for the SR-71 crews at a fraction of the cost of flying the Blackbirds. Subsonic, the T-38's flying characteristics, performance, and handling qualities replicated the SR-71 as close as any other aircraft.

The unit designation was changed to the 9th SRW on 25 June 1966. Colonel John DesPortes was the commander of the new wing while construction was taking place at Beale. He was soon promoted to brigadier general, after which Col. Doug Nelson took over the SR-71 wing. Nelson's immediate task would be to select a small group of highly competent commanders and SAC aviators to form the initial cadre of the SR-71 unit. He was busy visiting B-58 units, the U-2 organization, and Headquarters SAC to review the records of the best pilots and navigator/bombardiers among many highly qualified SAC aircrew members. From the beginning, SAC kept the SR-71 program in its own exclusive preserve, restricting recruitment to crew members from bombers and U-2s, with only a few exceptions. Steadily, Nelson found the finest officers in SAC for his new hand-picked unit. Two squadrons at Beale were formed for the arriving SR-71 crew members, called the 1st Strategic Reconnaissance Squadron (SRS) and the 99th SRS.

After passing the astronaut's physical exam at Brooks AFB, Texas, they arrived at Beale to begin preliminary training. Their security clearances were checked with a full background investigation. Once completed, they were briefed and cleared into the SR-71 program, called Senior Crown. Anyone who had access to the SR-71 program had to have a Senior Crown security clearance. Four of the B-58 crews undertook a seven-week training course on the SR-71 at the Skunk Works. They could spend their time with key Lockheed engineers, including Kelly Johnson and Ben Rich, and draft up lesson plans for future SR-71 training to be conducted at Beale AFB.

A HABU (PRONOUNCED "HAW BOO") IS A POISONOUS SNAKE found in Southeast Asia. Habus are pit vipers, more closely related to the adder. The actual habu is relatively small, usually not longer than five feet. It is not typically aggressive but will bite when provoked or backed into a corner. When the A-12s arrived on Okinawa in May of 1967, the locals thought this strange and somewhat evil-looking plane was similar to their black habu snake. Superstitious Okinawans pointed at the sinister black plane in the skies murmuring, "Habu . . . Habu." From then on the Habu became the nickname of the SR-71. (There is another version to the story of how the plane became named the Habu, but I tend to agree with this one.)

One tradition that was started by Chief Bill Gornik was the tie-cutting ceremony. Anyone who flew in the SR-71 Blackbird had to have their tie cut. When you flew your first operational sortie on Okinawa, you wore the ugliest tie you could find underneath the pressure suit. Upon landing the Det Commander would cut your tie in half.

Kelly Johnson had his tie ceremoniously cut in half by Chief Bill Gornik on the flight line. All the other maintenance troops helped with the ceremony. All the cut ties were hung from the striped pole. *Bill Gornik*

When the final crew list was settled it read as follows (pilot/RSO):

Crew 01: Maj. John Storrie and
Capt. Cosimo "Coz" Mallozzi
Crew 02: Maj. Robert "Gray" Sowers and
Maj. Butch Sheffield
Crew 03: Lt. Col. Al Hichew and
Maj. Tom Schmittou
Crew 04: Capt. Pete Collins and
Capt. Connie Seagroves
Crew 05: Maj. Jack Kennon and
Capt. Cecil Braden
Crew 06: Capt. Bill Campbell and
Capt. Al Pennington
Crew 07: Maj. Pat Halloran and
Capt. Mort Jarvis

Crew 08: Maj. Buddy Brown and
Capt. Dave Jensen
Crew 09: Capt. Dale Shelton and
Capt. Larry Boggess
Crew 10: Maj. Jerry O'Malley and
Capt. Ed Payne
Crew 11: Maj. Don Walbrecht and
Capt. Phil Loignon
Crew 12: Capt. Earl Boone and
Capt. Dewain Vick
Crew 13: Capt. Tony Bevacqua and
Capt. Jerry Crew

These were to be the first of many applicants wanting to fly the SR-71 Blackbird. They, and many others after them, developed future application

The bottom halves of all the operational crew members' ties were eventually stapled to a display board at Det 1, with name and mission date sewn onto each tie. The tie board and all squadron memorabilia now reside in the National Museum of the Air Force in Dayton, Ohio. *Rich Graham*

procedures, established the crew selection process, interview techniques, and the training program. Not knowing how much operational flying was in the future, Beale had two SR-71 squadrons from 1966 to 1971. At one point there were twenty-five pilots and twenty-five RSOs in the 1st SRS and the 99th SRS. The 99th SRS disbanded in March of 1971. To put this in perspective, when I arrived at Beale in June 1974 the 1st SRS had only nine operational crews.

SR-71 crews began a tradition of wearing a Habu patch on their flight suits after completion of their first operational reconnaissance mission on Okinawa. Upon landing, the pilot and RSO discovered a Habu patch mysteriously sewn on the left shoulder of their flight suits. Only those

The prized Habu patch was worn on the left shoulder of the flight suit. The "3 plus" patch was on the right shoulder. The 1st Strategic Reconnaissance Squadron patch was worn on the right front side of the flight suit. *Rich Graham*

who flew the SR-71 on operational missions could wear the highly coveted Habu patch. SR-71 crew members were also called Habus by other Air Force crew members.

CHAPTER 7

Selection Process

As an applicant, the SR-71 selection process seemed mystical. However, seen from the "inside," it was a very methodical screening program designed to choose the best crews with the aptitude to fly extremely sensitive reconnaissance mission all around the globe, over and around hostile foreign countries. Once the applicant put his "package" together—including all of his officer efficiency reports (OERs), standardization/evaluation check rides, endorsements, recommendations, biography, picture, flying time history, and any other relevant data—it was sent to the 9th Wing's Crew Selection Office. They were in charge of screening and maintaining the records of future candidates.

The application process allowed SR-71 crews to look over each applicant's file and make comments as to their suitability, especially if they knew the individual. First-hand endorsement from another crew member helped an applicant's chances of getting selected for the interview and physical. After that they were on their own. Whenever the SR-71 squadron commander determined a new crew would be needed, a selection board was convened. It typically met at least once every six months to select one or more crew positions needing to be filled about seven to nine months down the road.

The selection board typically comprised the director of operations (DO), the 1st SRS commander and operations officer, the chief of the Standardization/ Evaluation Branch (Stan/Eval), SR-71 simulator instructors, and other SR-71 pilots and RSOs having inputs. The pilot and RSO applicant folders were placed in two separate piles, which contained anywhere from five to ten applications for each position. Board members sat around the table and all the folders were passed out. After everyone had reviewed all the folders, each applicant was discussed. It was as fair and unbiased as a selection process could be. Selecting the "best of the best" was not easy.

In the end, the selection board chose a pilot and/or RSO to bring

Opposite: Roger Jacks' last flight. Seated, left to right, top to bottom: John Murphy, J. T. Vida, Don Emmons, Al Cirino, Tom Alison, John Fuller, Rich Graham, Buz Carpenter, Bill Groninger and Bruce Liebman. Standing at left: Bill Keller and, Chuck Sober (holding bottle). In pressure suits: Joe Kinego (left) and Roger Jacks. Standing at right: B. C. Thomas, Pat Bledsoe, who is almost obscured by the 9th Wing Commander, John Storrie. In front, left to right: Jim Sullivan, Jay Reid, and Tom Keck. *Rich Graham*

Right: Squadron pilots and wives get together in the squadron's "heritage room" for a solo party. Hanging on the wall are all the squadron crew members' names on brass plates and their crew photos. *Rich Graham*

The crew of Doug Soifer (left) and Mike Smith talk to a fourth grade class on Okinawa in February 1986. Many crews liked to volunteer in their local communities and overseas.
Doug Soifer

WHEN I BECAME PART of the interview process, my primary objective was to find out why the applicant wanted to apply for the SR-71 program. After posing a few probing questions, it usually became obvious to me if the applicant was using the high visibility of the SR-71 program as a "springboard" to help launch his Air Force career to greater heights. I had little patience for this type of mentality and in a heartbeat handed out a "Do Not Recommend" rating.

Over many years of selecting SR-71 crews, I came away with one observation: the more recommendations an applicant had from general officers, the *less* likely were his chances of getting into the program. The board was looking for endorsements from officers who had a first-hand knowledge of the applicant's abilities and airmanship skills—not how good he looked or how well he could push papers. Flying skills rather than politics was the name of the game. Occasionally, some general officer forced us to interview "their" candidate, in which case, we became very suspicious, particularly about how well he could fly an airplane. Few made it into our ranks.

out to Beale for a week of evaluation and interviews. During the week the applicant was scheduled for a personal interview with all the commanders, up to and including the wing commander. In actuality, he was also being evaluated by everyone associated with the SR-71 program. The uniqueness of the squadron required a strong bond between all crew members.

Sometimes, even the informal interview had dire consequences for the applicant. One applicant was at "Happy Hour" in the Beale officers club on the final night of his evaluation process.

LINK
SR-71
'MULATOR

All he had to do was be pleasant and he would have been hired. Instead, the old WWII British saying, "Loose lips sink battleships" came true. The applicant obviously drank more than he could hold and began to criticize the haircut standards of the other SR-71 crew members. He might have been right, but needless to say, he didn't make it into the program. Another applicant got so drunk and obnoxious at the officers' club he was not hired. Compatibility with fellow crew members was extremely important—a crew relied on each other to stay alive.

The goal was to select candidates who had something to contribute to the SR-71 program—not ones who were seeking something out of it. The formal interview was also a good time to let each applicant know about the extensive temporary duty (TDY) commitment SR-71 crews incurred. If the applicant was married, now was the time to be sure he understood that he wouldn't be seeing much of his family and to find out how he felt about it. After each formal interview a comment sheet was filled out on the applicant, as well as a recommendation to hire or not to hire.

The world's one-and-only SR-71 simulator on display at the Frontiers of Flight Museum in Dallas, Texas. The pilot and instructor pilot enter the simulator from the platform. The RSO cockpit is on the floor near the windows. The RSO instructor sits outside, in front of the two computer monitors. The brown cabinets house the simulator computers. A J-58 engine is in the foreground. *Rick Poole Photography*

Above: The instructor pilot's station in the simulator. The keyboard and computer screen at right is where he programs malfunctions and emergencies for the pilot to contend with. The station has repeater instruments overhead and on each side panel that display what the trainee sees. Once the trainee's ejection seat is locked in the full forward position (as seen here), three-axis motion of the simulator is enabled. *Rick Poole Photography*

Right: This is where the RSO instructor programs malfunctions for the RSO trainee to deal with. Both the pilot and RSO simulators can be run independently, allowing greater focus on individual problems. Once pilot and RSO proficiency were each sufficient, they were electronically linked together in the simulator as a crew. *Rick Poole Photography*

Each pilot applicant received a minimum of two evaluation flights in the T-38, with either the squadron commander or an SR-71 pilot qualified as an instructor pilot (IP) in the T-38. The applicant was evaluated on his general airmanship skills and his ability to stay ahead of the aircraft, a skill crucial to flying the SR-71 at over 2,000 mph. A T-38 evaluation form became part of the final selection board records.

Since the SR-71 simulator was such a vital part of the training program, the applicant flew it to evaluate how well he coped flying a new aircraft and whether he could handle the most basic flying skills without falling flat on his face. His application showed how well he performed in aircraft he had previously flown, but the selection board needed to observe how he managed under the stress of flying the SR-71. The simulator instructor would intentionally task-saturate

the prospective pilot or RSO, evaluating his capability to perform under stress in a completely foreign environment. For some candidates the SR-71 simulator was overwhelming, while others took things in stride. A completed simulator evaluation form was added to the paperwork.

One of the unspoken reasons that kept many Air Force aircrews from applying to the SR-71 program was the extensive physical examination. Originally, the applicants had to take the equivalent of an astronaut's physical at the Brooks Aerospace medical facility in San Antonio, Texas. In some instances, physical problems were discovered that permanently grounded several pilots. If you ever want to stop an Air Force pilot from doing something, just tell him it "might" permanently ground him from ever flying again. Once word got around about the extensive physical exam for the SR-71 program grounding applicants, aircrews became very reluctant to apply. It was either because they had something wrong with them that only they knew about, or they were afraid something would be discovered during the physical exam. In either case, many good aviators were not willing to take such a gamble.

Since the SR-71 program was under Strategic Air Command (SAC), many pilots would not apply, as that was an organization many did not want to be associated with for parochial reasons.

During the early 1970s, the SR-71 entrance physical exam was moderated to an expanded normal Air Force annual physical. It required an extra day at the hospital at Travis AFB, California, doing various tests that the Beale hospital was not equipped to handle. These included treadmill testing of the heart, an electroencephalogram (EEG) test of the brain, lung capacity test, glucose tolerance testing, extensive blood tests, full sinus X-rays, and a one hour interview with a "shrink."

During the interview the shrink asked a lot of basic questions about family, Air Force background, and why one wanted to fly the

SR-71. All of his probing was designed to put the applicant at ease and listen to him talk about things close and familiar to him, while the shrink was trying to determine if the applicant would be a sound and stable crew member. The program demanded level-headed crews that knew how to apply and follow very specific rules. It was important to weed out anyone who just might try something stupid or "off the wall" while flying highly sensitive reconnaissance missions over and around unfriendly countries. There is very little room for improvising at Mach 3+. The SR-71 squadron could not afford to have a rogue pilot put the entire program in jeopardy by ending up in international headlines. Unfortunately, one pilot slipped through the net and let his ego get the best of him while flying. He lied to the squadron commander and had to be thrown off the crew force in 1987.

Once the applicant departed Beale AFB, the final selection was often informal, and proceeded quickly, since by now there were only one or two prospective applicants at the most. The general discussion revolved around how well the applicant had performed. If the applicant was not accepted for SR-71 training, a very polite letter was sent to him from the squadron commander restating that the program was very, very competitive and to be chosen for the interview process was an honor in itself. Once the selection board voted "yes" on an applicant, he received a personal phone call from the squadron commander welcoming him into the program and established the timing for his training to begin. It was also a good time to let him know he faced a grueling ten-month training program that had no guarantees with it. There were still many hurdles to cross.

Depot-level maintenance for the SR-71 was accomplished at Lockheed's Plant 42 at Palmdale, California. Each plane was basically torn apart, inspected, and repaired. This required extensive industrial facilities, specialized tools and support equipment, and skilled personnel not available at lower levels of maintenance. *Lockheed Martin*

CHAPTER 8

Training Program

Most crews arrived for SR-71 training with an Air Force top secret (TS) security clearance. However, the SR-71 program was compartmentalized and demanded a special access required (SAR) security clearance called "Senior Crown." The Senior Crown security clearance was issued at Beale for all personnel who had a "need to know" about the SR-71 program and included practically everyone who worked on or with the Blackbird—from maintenance folks to the clerk typist in the squadron.

The SR-71 aircraft's operating manual (the "Dash-1") and pilot's checklist were classified "Secret Noforn." That security clearance meant the manuals were classified secret and had a "no foreign" government dissemination caveat. Once crews were briefed into the Senior Crown program, they signed for, and were issued, their classified flying manual and checklist at the squadron. To control all the classified documents, an entire wall in the squadron was lined with security safes to store the material. This was the crews' first introduction to an increasing maze of combination locks, cipher code, and secrets that they would have to memorize in order to fly a mission.

The first order of business before starting SR-71 training was to get checked out in the T-38 Talon aircraft, used as a companion trainer for SR-71 crew members. Another role for the T-38 was to fly "pace chase" formation with the SR-71. Whenever the SR-71 was in trouble and needed to be looked over externally, the T-38 was an ideal aircraft to do the job. Every time an SR-71 was flying at Beale, a T-38 had to be airborne or "cocked" on the ground, ready for immediate response with a qualified "pace chase" crew member onboard. When U-2 operations moved from Davis Monthan, Arizona, to Beale in 1975, the number of T-38s increased to fourteen aircraft to maintain the flying proficiency of both the U-2 and SR-71 pilots. The checkout in the T-38 took pilots about four to six weeks to complete all phases of flying: contact, instrument, formation, cross-country, and SR-71 pace chase.

I met my RSO, Don Emmons, for the first time in September 1974 during the flight line arrival ceremonies at Beale for Capt. Buck Adams and Maj. Bill Machoreck. They had just set the London-to-Los Angeles world record in the remarkable time of three hours forty-seven minutes, thirty-nine seconds (5,463 miles) while returning from the Farnborough Air Show.

Opposite: The SR-71B model trainer 956 in which I flew U.S. Senator Howard Cannon on a distinguished visitor (DV) flight in 1977. Typically, the DV sat in the back seat with the instructor pilot in the front because several switches in the front cockpit are not duplicated in the rear. *Tony Landis*

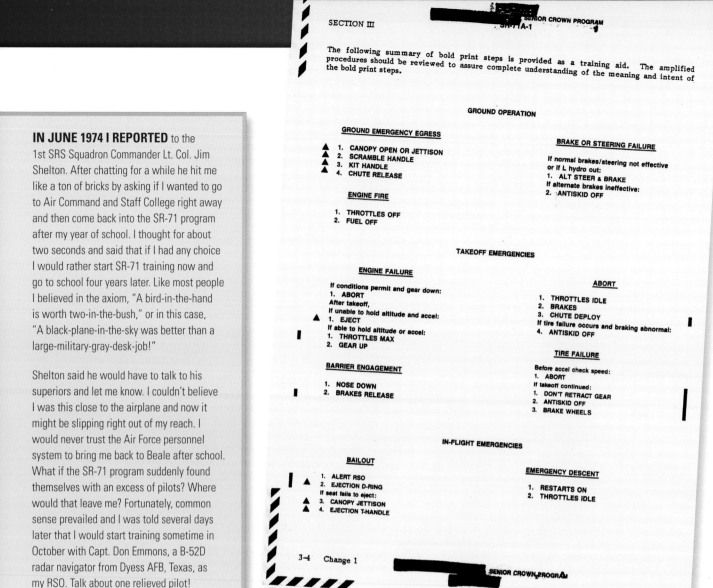

SECTION III

The following summary of bold print steps is provided as a training aid. The amplified procedures should be reviewed to assure complete understanding of the meaning and intent of the bold print steps.

GROUND OPERATION

GROUND EMERGENCY EGRESS

1. CANOPY OPEN OR JETTISON
2. SCRAMBLE HANDLE
3. KIT HANDLE
4. CHUTE RELEASE

BRAKE OR STEERING FAILURE

If normal brakes/steering not effective or if L hydro out:
1. ALT STEER & BRAKE
If alternate brakes ineffective:
2. ANTISKID OFF

ENGINE FIRE

1. THROTTLES OFF
2. FUEL OFF

TAKEOFF EMERGENCIES

ENGINE FAILURE

If conditions permit and gear down:
1. ABORT
After takeoff,
If unable to hold altitude and accel:
1. EJECT
If able to hold altitude or accel:
1. THROTTLES MAX
2. GEAR UP

ABORT

1. THROTTLES IDLE
2. BRAKES
3. CHUTE DEPLOY
If tire failure occurs and braking abnormal:
4. ANTISKID OFF

TIRE FAILURE

Before accel check speed:
1. ABORT
If takeoff continued:
1. DON'T RETRACT GEAR
2. ANTISKID OFF
3. BRAKE WHEELS

BARRIER ENGAGEMENT

1. NOSE DOWN
2. BRAKES RELEASE

IN-FLIGHT EMERGENCIES

BAILOUT

1. ALERT RSO
2. EJECTION D-RING
If seat fails to eject:
3. CANOPY JETTISON
4. EJECTION T-HANDLE

EMERGENCY DESCENT

1. RESTARTS ON
2. THROTTLES IDLE

3-4   Change 1

SENIOR CROWN PROGRAM

IN JUNE 1974 I REPORTED to the 1st SRS Squadron Commander Lt. Col. Jim Shelton. After chatting for a while he hit me like a ton of bricks by asking if I wanted to go to Air Command and Staff College right away and then come back into the SR-71 program after my year of school. I thought for about two seconds and said that if I had any choice I would rather start SR-71 training now and go to school four years later. Like most people I believed in the axiom, "A bird-in-the-hand is worth two-in-the-bush," or in this case, "A black-plane-in-the-sky was better than a large-military-gray-desk-job!"

Shelton said he would have to talk to his superiors and let me know. I couldn't believe I was this close to the airplane and now it might be slipping right out of my reach. I would never trust the Air Force personnel system to bring me back to Beale after school. What if the SR-71 program suddenly found themselves with an excess of pilots? Where would that leave me? Fortunately, common sense prevailed and I was told several days later that I would start training sometime in October with Capt. Don Emmons, a B-52D radar navigator from Dyess AFB, Texas, as my RSO. Talk about one relieved pilot!

Taken from the SR-71 flight manual are various emergencies actions that have to be not only memorized, but practiced in the simulator. There were a total of sixty bold print steps that had to be committed to memory... and if you flew the T-38 as well, those had to be memorized and not confused with the SR-71's emergency steps! The word SECRET was crossed out once the program was declassified in 1990. *Rich Graham*

SR-71 training lasted about ten months. The goal was to complete a hundred hours in the aircraft, after which crews were considered combat ready and cleared to fly operational sorties around the globe. The hundred-hour requirement ensured a certain level of proficiency in flying various mission profiles anywhere in the world.

Our training started in mid-October. The SR-71 training program was not an established Air Force school. It had no formal guidance from the Air Force, nor any assigned ground or flight instructors. All the training was accomplished by regular SR-71 crew members. For the first two weeks, aircrews attended a field training detachment (FTD) course, designed to teach new aircraft mechanics in the "nuts and bolts" of maintaining and repairing the plane. For crews it provided a solid background and overview of all the aircraft's systems. Upon completion of the FTD course, intensive simulator training began.

The simulator training was both demanding and stressful—every time you stepped into the simulator as a student you sweated bullets. Most crews received around 120 hours flying simulator missions before ever putting a foot inside the actual aircraft. All the simulator missions were well thought out and included an outline and accompanying lesson plan. Each training mission became increasingly more complex, building on the previous

ones. Their scenarios were developed and refined over the years by crews who found a better way of presenting and instructing the training.

Crews had unrestricted access to the simulator, enabling them to study and prepare for each simulator mission at their own pace. Each simulator mission lasted about four hours. A typical simulator period had the pilot and RSO arriving early so they could go over any last minute details and have all the mission materials laid out and ready before the instructors arrived. The instructor pilot briefed the overall mission profile and emergencies to be covered during the session.

It may sound easy, but one of the most difficult tasks for new pilots to learn was to be able to accurately describe what problems were occurring in his cockpit so that the RSO could understand, locate, and read the correct checklist procedure. Many experienced pilots have misidentified the aircraft's real problem(s) and had the RSO reading a similar, but incorrect, checklist procedure. That's precisely why the RSOs needed an excellent working knowledge of the entire aircraft and its systems. Crew

coordination was an area where crews either passed or failed miserably! The simulator program developed a trust and bond between each other that allowed two separate individuals, sitting in separate cockpits, with two different roles, to perform like one individual, accomplishing a very demanding job.

Former SR–71 pilot Bernie Smith tells an exciting story that exemplifies that close relationship with his RSO, E. D. McKim.

On 22 December 1982, about one and a half hours into a night mission while eastbound over Canada, we experienced a left generator failure. We selected

Above: Capt. Buck Adams followed by Maj. Bill Machoreck come down the ladder at Beale AFB after completing their speed run from London to Los Angeles. The civilian with the trumpet is Bill Brown, one of the tech reps, playing his traditional "Charge" at major SR-71 events. *USAF*

Left: SR-71 pilot Jim Sullivan (right) and RSO Noel Widdifield at the microphone in front of a large international press group at the completion of their record-breaking speed run from New York to London. *Lockheed Martin*

ON ONE OCCASION I needed the "pace chase" aircraft to look me over. It was right after takeoff, when my RSO, Don Emmons, and I heard a loud "bang" coming from somewhere in front of the rudder pedals. I was approaching our climb out speed of 400 knots and thought I might have hit a bird. Looking out over the nose, I noticed a large piece of the nose section beneath the aircraft had blown open, still attached, but flapping in the air stream. I couldn't tell the full extent of the

Left: The removable nose section (here on a maintenance dolly) allowed the aircraft to have radar-imaging capability with the advanced synthetic aperture radar system (ASARS), photographic imagery with the optical bar camera (OBC), or a ballast nose installed. The nose section was held on by four massive fasteners. *Rich Graham*

damage because the piece was underneath the nose and all I could see was one end of it flapping about. Joe Kinego, another SR-71 crew member, remembered the "funny (then scary)" radio call Don made back to Beale informing everyone of our problem: "The nose is coming off!" That radio call got everyone's attention!

The T-38 chase aircraft joined up in formation and told us the extent of our damage as we returned to Beale. It was comforting to know everything else was intact. A one-by-two foot metal section of the nose had ripped open while still attached underneath. I always felt fortunate that it happened during daylight. Had it been at night, I'm positive I wouldn't have been able to see the loose panel and might have discounted the possibility of anything being wrong and continued the mission . . . you just never know.

Talking to Lockheed engineers later, they determined there was a good possibility if we had gone supersonic, the entire nose section (which is removable) might have imploded from the supersonic shock wave. That scenario would have probably necessitated an ejection for both of us!

Below: A T-38 crew practices flying chase on the SR-71 in the Beale training area. The T-38 is much more maneuverable than the SR-71, but has to be careful not to fly directly behind the Blackbird. The tail art on 974 is a Habu snake entwined around the number "1" for Det 1. The words "ichi ban" (below the snake) translate from Japanese into "number one." *Lockheed Martin*

Airplane Arrangement

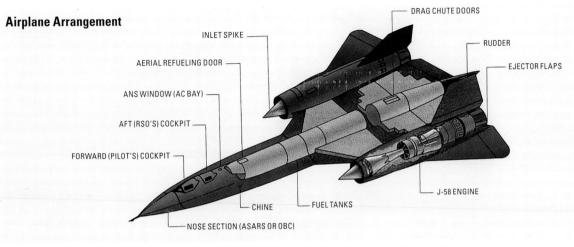

DRAG CHUTE DOORS
INLET SPIKE
RUDDER
AERIAL REFUELING DOOR
EJECTOR FLAPS
ANS WINDOW (AC BAY)
AFT (RSO'S) COCKPIT
FORWARD (PILOT'S) COCKPIT
J-58 ENGINE
CHINE
FUEL TANKS
NOSE SECTION (ASARS OR OBC)

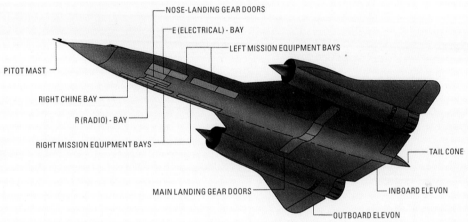

NOSE-LANDING GEAR DOORS
E (ELECTRICAL) - BAY
LEFT MISSION EQUIPMENT BAYS
PITOT MAST
RIGHT CHINE BAY
R (RADIO) - BAY
RIGHT MISSION EQUIPMENT BAYS
TAIL CONE
MAIN LANDING GEAR DOORS
INBOARD ELEVON
OUTBOARD ELEVON

Grand Forks AFB as the nearest suitable emergency recovery airfield. During the descent the primary hydraulic system failed, indicated an accessory drive system (ADS) failure. An ADS failure is one of the most serious emergencies that can occur on the SR-71. Landing became imperative.

We dumped fuel to reduce our landing gross weight as we neared Grand Forks. The weather was deteriorating rapidly. It was now 200 feet overcast, one-half mile visibility, and a slippery runway with freezing drizzle. During the emergency I flew the ILS approach down to minimums. At minimums you should execute a missed approach if the runway environment is not in sight. The visibility was so bad I could not see anything. Since there were no suitable alternates

available and since we were low on fuel, the idea of a missed approach was not too exciting. I'll never forget E.D.'s word as I sat there at minimums, staring into an opaque windshield, "I have the strobe lights insight, you're on the centerline, keep on coming." With no forward visibility from the back seat, he was looking through his View Sight and had picked up the Sequence Flashing Lights beneath the aircraft. It was this bond, or trust, I had in him that made me pull the throttle to idle and sit there while I felt the runway touchdown.

The simulator missions were numbered one through eleven, each one marking a major milestone before you could move on

The six main fuel tanks (blue) held 80,250 pounds of fuel. Transferring fuel between the most forward and aft tanks controlled the center of gravity (CG). The chine (from the nose to the start of the leading edge of the wings) provided a substantial portion of the total lift at high supersonic speeds and eliminated a need for canard surfaces. *Lockheed Martin*

The RSO's simulator cockpit. Notice there are no flight controls. The only outside view the RSO has are the two large windows on each side and an electro-optical viewsight (top center glass screen), allowing him to look beneath the aircraft. *Rick Poole Photography*

to the next numbered mission. Simulator mission number twelve was the final check ride. Before each numbered mission, the student crew received a practice mission covering the same material as the numbered mission. If they successfully passed the practice mission, they were recommended for the numbered mission. If a crew failed a numbered mission, they received extra simulators and additional help preparing them for the recheck. A recheck was usually given by the most experienced SR-71 crew at Beale to make sure the student crew performed up to accepted standards, and to add credibility to their program continuation recommendation in case the student crew did not perform well.

How badly the student crew screwed up, and what this involved, determined the fate of the crew. They could either receive additional training or be released from the SR-71 program. I am only aware of three pilots and one RSO eliminated from the SR-71 program. The pilots were all skilled aviators, but just not cut out to be SR-71 pilots. All three went back to flying their former aircraft; one of them was later killed when his RF-4 crashed. The RSO who failed was "sponsored" into the program by a four-star general officer. He hadn't flown recently, had minimum flying time, and didn't perform particularly

well on the interview, but was "issued" to us anyway. He was eventually eliminated from the program.

After completion of simulator mission number twelve, the crew was ready to begin flight training in the SR-71. At one time there were two B-model trainer aircraft (tail numbers 956 and 957). Unfortunately, on 11 January 1969 while on a training flight, an emergency in aircraft 957 caused both the instructor and student pilot to bailout seven miles short of Beale's runway. The plane was destroyed in the crash. It was obvious if the only remaining trainer crashed, all SR-71 training would come to an immediate halt. Another SR-71 trainer model variation, the SR-71C, was quickly built, and first flew on 14 March 1969. This C-model trainer (tail number 981) was unique: It was a hybrid, consisting of the forward fuselage from a static test specimen at the Lockheed Skunk Works, mated to the wing and rear section of a formerly crashed YF-12. Maintenance personnel nicknamed it "The Bastard" because of its origin, and the fact that standard technical data for maintaining the plane did not exist.

The pilot training, dual-seated Blackbirds did not carry any reconnaissance sensors. They were only equipped to teach the new pilots how to fly and refuel the SR-71. Every new pilot received five training flights with

The pilot's cockpit. The black panels at the top left and right side were called "bat wings" and used to block out the sun. The sun is so brilliant at 80,000 feet that it can "wash out" the instruments to where they become unreadable. *Rick Poole Photography*

This static test specimen was joined with the salvaged aft section of a crashed YF-12 to produce the SR-71C model trainer, tail number 981. When I was going through my training in 1974, I flew the SR-71C model on two of my five training sorties. *Lockheed Martin*

YF-12 934 has high-speed camera pods mounted under each engine nacelle, used to record the launch and separation of the AIM-47 missile. After a crash-landing, its tail section was mated with the static test specimen to become the second trainer. *Lockheed Martin*

AFTER YOUR FIRST FLIGHT in the SR-71 at Mach 3+, each crew member was presented with a highly prized Mach 3/SR-71 lapel pin, a Lockheed SR-71 model plane, a Mach 3 certificate suitable for framing, and a Mach 3 wallet card signed by C. L. "Kelly" Johnson. We often joked that you could probably get a free cup of coffee anywhere just by presenting the Mach 3 card.

Author's prized "Mach Three Plus Club" card featuring Kelly Johnson's actual signature. If you're in the presence of other crew members and don't have your card, it will cost you a round of drinks! *Rich Graham*

For Don and I, however, the Mach 3 card proved extremely valuable at the Farnborough International Air Show, England, in September of 1976. Having a few days off from work at Mildenhall, we drove to the village of Farnborough and found ourselves a bed and breakfast to stay at while attending the air show. After walking around all the exhibits on the first day, we passed by the Lockheed chalet and asked the security guard if we could possibly go inside because of our association with Lockheed's SR-71. He said, "Let me check" and disappeared inside.

He came back shortly with the Lockheed representative in charge of the chalet, who asked if we could prove our SR-71 affiliation. Immediately, Don and I whipped out our Mach 3 wallet certificates signed by Kelly Johnson. He was visibly impressed

and escorted us inside the chalet. It was only two years earlier that the SR-71 set the world speed record between New York and London and landed at the Farnborough Air Show for static display. That historic event was still fresh in aviation circles. He escorted Don and I to the bar for drinks and then showed us to the buffet table. We chatted to several Lockheed employees, and once the word spread that two SR-71 crew members were there, everyone wanted to meet us. We felt like rock stars over our five minutes of fame! Over the next few days we didn't have to buy any food or drinks and had the best seat in the house to watch the air show. To this day I still carry my Mach 3+ wallet card, hoping for another free admission!

Don and I once created a sonic boom that neither we nor the mission planners had anticipated. During our descent into the Andrews AFB annual air show in July of 1976, we inadvertently boomed a small country town in West Virginia at low altitude and created a real panic. The town's only industry was a dynamite manufacturing plant. When everyone in town heard the long dreaded "BOOM!!," they were convinced the factory had blown up. Their volunteer fire department, along with the rest of the town's folks, raced out to the factory only to find it still intact. A local newspaper clipping, describing the false alarm, was later sent to Don and me at Beale.

A practice single-engine go-around. The pilot uses the afterburner for thrust and, with the J-58 engines displaced far left and right of the aircraft's centerline, it required a large amount of left rudder to keep the plane flying straight ahead, as clearly visible. Refueling with one engine was an art requiring a large amount of bank and yaw into the good engine to stay on the refueling boom. *Lockheed Martin*

an instructor pilot in the rear cockpit of the SR-71B/C model. After the five flights, the new pilot flew his first "solo" with an experienced RSO in the SR-71. Every new RSO received two flights in an SR-71 with a seasoned pilot before flying solo with his own appointed student pilot.

From that point onward, the pace and level of mission sophistication increased in the SR-71, while the simulator training slowly decreased. Crews practiced high Mach and/or high-banked training missions. For flying in tight geographical confines, crews practiced low Mach (2.8 and 2.6) flying with high-banked missions. Normal banked Mach 3.0 missions were between thirty- to thirty-five degrees of bank. Steep banked turns were forty-five degrees maximum. Once a crew reached forty-nine degrees of bank at Mach 3.0, they were at the structural limits of the aircraft for a level turn. They also underwent night flight training. The goal was to be confident and proficient in handling the SR-71 in any worldwide scenario. Once the aircrew achieved a

hundred hours of flying time, they were certified combat ready to fly their first operational missions out of Okinawa.

It was difficult to develop training routes in the United States because of the sonic boom created by the shock wave coming off the airplane at supersonic speeds. At 80,000 feet the sonic boom could be heard on the ground as a double thunder clap, the first clap much louder than the second. Some of the major obstacles our routes had to avoid were: cities in excess of thirty thousand people, all national parks and recreation areas, national monuments, and winter skiing resorts. There were also individual locations the aircraft had to avoid where the sonic boom supposedly created problems for farmers. Mink farmers said their minks wouldn't mate and farmers claimed their cows wouldn't give milk or chickens wouldn't lay eggs, all because of the sonic boom. Beale AFB had a full-time officer responsible for investigating all sonic boom complaints. The question remains if any of these complaints carried any merit!

On 11 January 1968, instructor pilot Gray Sowers and student pilot Dave Fruehauf were flying a training sortie at Beale AFB in the B-model aircraft 957. Dual generator failure caused the fuel pumps to become inoperative, and both engines eventually quit seven miles north of Beale. Both men ejected successfully. *USAF*

CHAPTER 9

Physiological Support Division and the Pressure Suit

Space begins at about 125 miles above the earth, but as a physiological environment it begins at around 63,000 feet, where the atmospheric pressure becomes so low that fluids boil at the body temperature of 98.6 degrees Fahrenheit (F). The main function of the pressure suit was to save your life at the extreme altitudes, temperatures, and speed at which the SR-71 flew. Above 45,000 feet any crew member's effective performance time (EPT) is between nine to twelve seconds without oxygen. EPT is the amount of time an individual is able to perform useful flying duties in the absence of oxygen before going unconscious. Without a pressure suit at 80,000 feet, and the loss of all cabin pressure or an ejection scenario, you could not survive.

The organization at Beale that maintained the pressure suits was the Physiological Support Division (PSD). The facility was located on the flight line and was the Air Force's entire repository for all pressure suit operations. Consequently, PSD had a high level of experienced personnel, who had the technical expertise and capability to do a complete overhaul on pressure suits. To SR-71 crew members, they were highly qualified "technicians" in every sense of the word.

Although not considered a military uniform, the David Clark Company Model 1030 pressure suit was the most prized uniform for aspiring Habus. It meant you were one step closer to flying the aircraft! Before donning the pressure suit each crew member was given a locker in the PSD changing

Opposite: The pressure suit and helmet weighed around fifty pounds and cost just under $200,000. The portable liquid oxygen converter the pilot is carrying supplies cooling air to the pressure suit until connected to the aircraft cooling system. *Lockheed Martin*

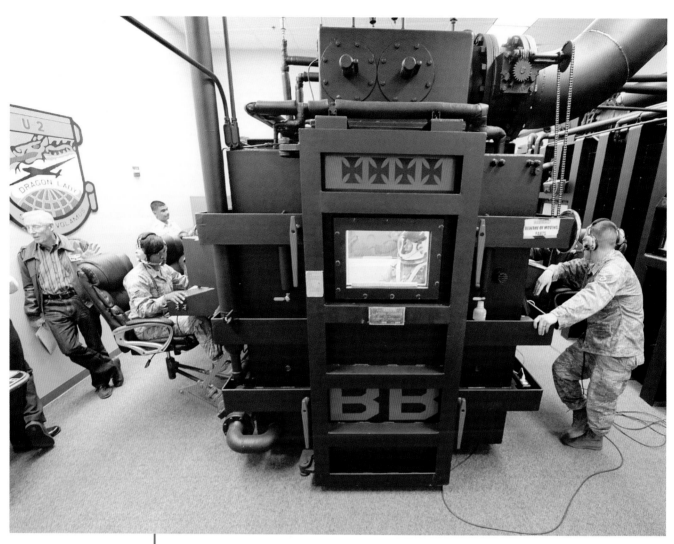

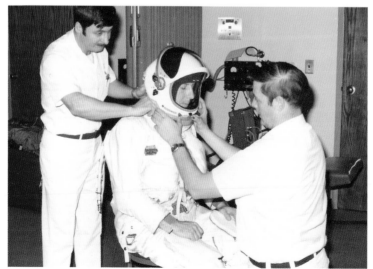

Above: The altitude chamber at Beale AFB allows crew members to experience high altitude physiological problems in a very controlled environment. *USAF*

Right: The author suiting up in the white version of the pressure suit. The original suits were silver, then changed to the white Nomex (flame resistant) exterior. When I arrived at Beale in 1974, I wore the white pressure suit and then changed to a dark brown exterior in 1976. Around 1978 they changed to the yellow-gold color, which remains in use today by U-2 pilots. *Rich Graham*

room to store personal effects and their military uniform before flying. Each crew member was issued four pairs of cotton, long underwear (very sexy!), and thick white socks to be worn under the pressure suit. The underwear provided a layer of warmth and reduced skin irritation from the inner liner of the pressure suit.

The Model 1030 pressure suit came in twelve basic sizes. From there everyone was individually fitted with adjustments by arm length, glove size, and foot size. Although the pressure suit was a six-layer outfit, three layers were significant. The outermost layer (exterior) was made of fire-retardant material called Nomex. It contained zippered pockets on the upper and lower legs, Velcro patches on the upper legs to secure checklists, and, most importantly, the parachute harness connections.

The inner layer, called the bladder, was made of rubber compound, and became inflated, much like a balloon, when air

The Lockheed SR-1 ejection seat used in the SR-71 was very reliable for ejecting from zero speed and altitude to Mach 3+ and 90,000 feet. The large yellow D-ring between the leg padding initiates ejection and is also a handhold for protection of the arms during ejection. One quick, hard tug on the D-ring and you're on the way out of the aircraft. *Tony Landis*

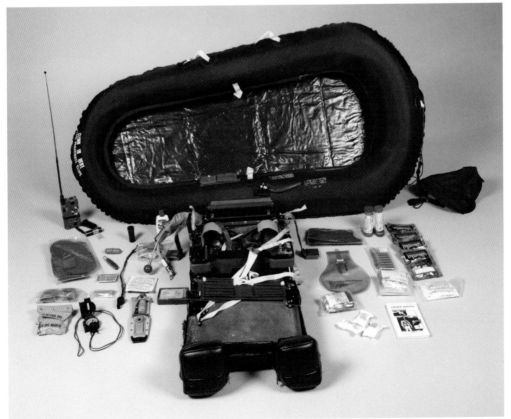

A survival kit is located underneath the seat cushion. It contains emergency oxygen supply and basic survival equipment for landing on the ground or in water. The one-man raft inflates automatically once the survival kit handle is pulled. Every crew member practiced a water landing in the pressure suit at the Air Force water survival course in Florida by being towed up to several hundred feet in a parachute and then released. *NASA*

Above: After rapid decompression, fogging inside the chamber dissipates and the pressure suit is fully inflated. As the suit swells, the helmet neck ring is kept from rising up and over your face by an adjustable nylon strap connected to a wire that goes around both sides of the neck ring, holding it in place. *USAF*

Right: PSD technicians control a simulated climb to 80,000 feet in the chamber and talk to the crew members about what they are experiencing throughout. On the other side of the chamber, additional technicians monitor progress. *USAF*

pressure was added to the suit. The rubber layer was irritating to bare skin, thus the need for a comfort liner made from lightweight Dacron material. Also located inside the bladder layer was a network of tubes to direct cooling air throughout the suit, particularly to the extremities.

Between the outer layer and the bladder of the suit was a tightly woven mesh netting designed to keep the bladder from inflating too much. The netting was woven in a special manner so that increased pressure within the inner lining caused the webbing to grow more rigid and contain the expansion.

The PSD facility also housed an altitude chamber capable of evacuating the air to an altitude of 85,000 feet. The purpose of the small chamber was to expose new crew members to the mating of the pressure suit to the ejection seat, build their confidence in the suit, teach them how to operate and control the suit's temperature and pressure, and, finally, to experience a loss of all cabin pressure at 85,000 feet. This was called a rapid decompression, since it happened in an instant.

Crews put their lives into the hands of the PSD technicians every time they flew. Wearing the pressure suit made it impossible to strap oneself into the SR-71's ejection seat making all the necessary connections. Consequently, crews were taught to extend their arms out each side of the cockpit and

Captain William D. Stanford helps actor Craig T. Nelson with the controls on his pressure suit at Beale AFB. Nelson, who portrayed Colonel Sarnac on the TV series *Call to Glory*, is preparing to shoot episodes depicting him as an SR-71 pilot. The series debuted on 13 August 1984. *USAF*

Crew members do not practice the physical ejection from an aircraft, but they do train for the parachute descent with a pressure suit harness on. Being able to operate all the survival equipment that comes down with the parachute is also part of the training. *USAF*

sit there patiently while PSD technicians, on both sides of the cockpit, mated the pressure suit to the aircraft and its ejection seat. They were extremely professional and safety conscious at their job, alert at all times to the danger of being complacent.

Once the crew member was strapped in the ejection seat inside the altitude chamber, the technicians departed and closed the thick door leading out of the chamber. After the chamber was fully sealed, they began evacuating the air pressure inside. As the chamber reached around 25,000 feet, the technicians asked the crew member over the interphone if he was experiencing any sinus problems. If not, the steady climb to 85,000 feet continued. The first thing you noticed on the climb happened at 63,000 feet. A flask of warm water (98.6 degrees F) placed inside the chamber slowly began to boil. By the time the chamber reached 70,000 feet, it was boiling rapidly, and at 80,000 feet the water was all gone—evaporated! The water was placed in the chamber to give you an appreciation of what could theoretically happen to your blood or other fluids without a pressure suit on.

To prepare the chamber for the rapid decompression, the greatly reduced air pressure at 85,000 feet was captured and stored in the larger altitude chamber, connected to the small chamber by huge air valves. Once the thin air was captured, they brought the small chamber back down to 26,000 feet. This was the normal cabin pressure inside the SR-71's cockpits, flying at 85,000 feet. Since the RD happened so rapidly, the supervisor briefed you on what to expect before throwing the decompression switch: "There will be a loud, explosive 'BANG' accompanied by immediate fogging, and then rapid clearing inside the chamber. Simultaneously, the pressure suit will fully inflate and become very rigid."

When he moved the switch, instantly the 26,000-foot cabin pressure would be at 85,000 feet. Just as advertised came the bang, fogging and clearing, and full suit inflation. To practice various cockpit chores with a fully inflated suit, they kept the altitude at 85,000 feet. It is surprising just how difficult it was to turn your head inside the inflated pressure suit. The most important task was to be able to reach for the ejection seat handle located between your legs and pull it upward to initiate the ejection sequence. Hopefully, crews would never have to experience an RD for real, but for now they had an ideal opportunity to experience one in a controlled environment.

CHAPTER 10

Command and Control

The vast majority of operational reconnaissance sorties were flown from either Kadena or RAF Mildenhall in Suffolk, England. At each of these locations were detachments ("Det" for short) of the 9th SRW at Beale. Over the years, the Det on Okinawa changed its name several times but finally settled down as Det 1 in 1974. Det 4 was the designation given to the unit at Mildenhall. Each Det was commanded by a former SR-71 crew member (with one exception) with the rank of lieutenant colonel or colonel. Although each Det was completely autonomous, both were supported by the host base for various needs such as housing, transportation, and dining facilities. Det 1 was always manned by Air Force personnel, with a small contingent of Lockheed technical representatives ("tech reps") and various other civilian contractors to help maintain the SR-71's unique sensors. By contrast, Det 4 originally used Air Force maintenance personnel, but in the early 1980s converted to a civilian Lockheed contract maintenance program. Det 1 had around three hundred personnel, Det 4 around two hundred.

Early on, personnel at each Det were assigned on a TDY basis from Beale. As each Det grew in responsibility, and as it appeared the SR-71s would be there for the long haul, maintenance and support personnel were assigned to the Dets on a normal Air Force overseas tour of duty. About the only personnel who continued going to the Dets on a TDY basis were the SR-71 crews, KC-135Q crews, and the PSD personnel who maintained our pressure suits. Each Det commander "owned" the SR-71 crews and PSD personnel while assigned there.

Considered a national asset and designed for strategic reconnaissance, the SR-71 required National Command Authority (NCA) approval for peacetime tasking. To avoid repetition of the unfortunate Francis Gary Powers U-2 incident, established procedures for SR-71 peacetime tasking became involved and complicated.

Opposite: SR-71 958 flying over the rice paddies and rich farming country surrounding Beale AFB in northern California. *Lockheed Martin*

Over the years, various "tail art" has been seen on the SR-71's rudders. The dart sticking in the double two represents Det 4, so named because it was the fourth detachment assigned to the 9th Wing at Beale. *Lockheed Martin*

A theater commander in chief (CINC) directed his request for SR-71 coverage to the Pentagon's Joint Reconnaissance Center (JRC). There, the "book officer" analyzed the request to find out if other Air Force assets were already covering the target and if the request was appropriate for the SR-71. If satisfied, he then coordinated with the Defense Intelligence Agency (DIA) to ensure there was no redundant coverage by sources outside the Air Force. If the DIA "passed" on the request, the book officer briefed the request up through the chain of command to representatives from the Joint Chiefs of Staff (JCS), the Secretary of Defense (SECDEF), and the president's national security advisor, who could approve the request in the name of the president. This process took about two weeks.

Former SR-71 crew members worked in the JRC as book officers and mission monitors for all operational sorties. The JRC also worked diplomatic issues. If overflight or landing rights had to be secured from a friendly country, it was the JRC's responsibility to obtain approval from the State Department. For every SR-71 operational mission flown, the JRC developed a threat matrix, weighing the political and military risk of flying the sortie against the needs of gathering the intelligence. Any direct overflight of a foreign country required specific presidential approval.

The JRC then notified the Strategic Reconnaissance Center (SRC) at Offutt AFB, Nebraska, of the requirement. Created in 1965 to provide immediate response to the requests of national intelligence users, the SRC's control center operated twenty-four hours a day and monitored all of SAC's operational reconnaissance missions while they were in progress. The SRC evaluated possible threats to the aircraft; coordinated the movement of aircraft, sensors, and support staff and equipment; and scheduled the sortie. The SRC maintained operational control over every reconnaissance mission flown by the SR-71.

Well in advance, the JCS sent the SRC a monthly schedule that included the number of sorties approved. SRC then determined the precise day and time the sorties would launch, taking into account such variables as weather, maintenance, tanker logistics, and other support requirements. The commander in chief, SAC (CINCSAC), and his senior staff reviewed the monthly and daily reconnaissance schedules and approved every mission prior to execution. The tasking for each operational mission came from SAC headquarters via secure communications and was addressed to all key players and interested parties throughout the world who had the need to know. Just hours prior to every mission, SAC sent the "execute" (go/no-go) message to the appropriate Det. That was the final approval needed to fly the mission. In case there were any last minute issues or questions, each Det had secure phones to contact SRC or anyone else in the chain of command and control.

In the Pentagon the office responsible for funding the SR-71 program, as well as other reconnaissance assets, had the office symbol of XOORZ. A staff officer from XOORZ briefed every operational sortie up the Air Force chain of command, including the secretary of the Air Force.

One of the biggest myths surrounding the SR-71 reconnaissance program is that the

DURING THE TIME I WAS THE 9TH WING COMMANDER at Beale, there were five detachments (Dets) around the globe. Each detachment had around 150–300 personnel to operate and maintain their aircraft. Det-1 flew the SR-71s located at Kadena Air Base, Okinawa, Japan; Det-2 flew the U-2 from Osan Air Base, South Korea; Det-3 flew the U-2 at RAF Akrotiri, Cyprus; Det-4 flew the SR-71 out of RAF Mildenhall, England; and Det-5 flew the U-2 out of Patrick AFB, Florida.

Having selected each Det commander, I knew their capabilities and entrusted each of them to run their own operation. However, the 9th Wing Commander at Beale was ultimately responsible for all the U-2, SR-71, and KC-135Q operations around the world. I had only one rule with each Det commander. I wanted to be informed of any event that could possibly escalate to the general officer ranks. I didn't want to be blindsided by a phone call from some general without knowing the answer first. In my office, and at my house on Beale, I had the ability to call, or be called by, any of the Det commanders. As you can well imagine, with the global operations, my phone rang anytime, day or night. My wife soon learned to sleep through a midnight call!

The SR-71 Squadron Commanders' primary job focused on the crew members and their families. At Beale the squadron commander had to ensure the right crew members were selected for the SR-71 program and trained to a high standard.

All of the 9th Strategic Reconnaissance Wing's aircraft. In 1987 the 9th Wing Commander had thirty-five KC-135Q tankers, fourteen T-38s, and all the SR-71 and U-2 aircraft under his command and control to fulfill the global reconnaissance mission at five detachments. *USAF*

plane has overflown the USSR and China. In truth, neither the A-12 nor the SR-71 has ever overflown the landmass of the USSR or China. After Gary Francis Powers was shot down on 1 May 1960 over Russia, no U.S. president would authorize direct overflights of the two superpowers. Name any other foreign country that the United States had an interest in, and the SR-71 has probably flown over it, as well as covering the periphery of the USSR and China while in international airspace. Yes, the sensors were capable of gathering intelligence deep inside these countries.

Mission planners at the Dets prepared the flight track, planned the mission, and

Names and an American flag patch were attached to the pressure suits with Velcro on the left shoulder. In case of capture by a foreign country, the aircraft markings, Geneva Convention card, and American flag patch were proof positive we were Air Force pilots. Sewn on the left and right lower sleeve were checklists of what to accomplish on the way down in the parachute after an ejection from the plane. *Lockheed Martin*

briefed the pilot and RSO. All SR-71 sorties around the USSR and China were flown under the Peacetime Aerial Reconnaissance Program (PARPRO) rules. These flights were flown in international airspace under VFR (visual flight rules) conditions and avoided threats or denied territory. If, for any reason, the crew could not maintain the PARPRO restrictions, they aborted the mission and returned to base. Sometime after the SR-71 program terminated, PARPRO was later changed to a new, all-encompassing acronym: PRCSO (Peacetime Reconnaissance and Certain Sensitive Operations).

One question that comes up routinely when talking about the SR-71 is, "What's it like spying on another country with the SR-71?" Habus had to be prepared with an answer rather than avoid the question. The point is, Habus were *not* spying. Covert spying on foreign countries is highly illegal, and people can be tried, convicted, and imprisoned for the act. By the internationally agreed-upon PARPRO rules, the SR-71 gathered intelligence *overtly* on foreign countries. For that reason, and by the Hague Convention of 1907, crew members wore the American flag on the left shoulder of the pressure suit with

name and rank included, as well as carrying Geneva Convention cards. All the SR-71s were painted with Air Force markings. If we ever found ourselves on the ground in a hostile country, there was to be no doubt we were Air Force crews flying an Air Force aircraft.

The majority of our operational sorties were flown repeatedly, with only minor deviations to the track or sensor operation. Most of these "routine" missions were flown to gather what was called "indications and warning" (I & W) intelligence. Flying over the same area on a regular basis allowed intelligence analysts to determine such things as troop movements, changes to the electronic order of battle (EOB), and aircraft deployments—all important indications and warnings that a problem was arising. If the I & W intelligence indicated an increased level of activity, mission planners focused the SR-71's intelligence-gathering sensors in greater detail and frequency on those particular targets.

Several missions at each Det had a higher priority than others and, as such, included several means of insuring they would be flown successfully. The lowest level of priority had a spare SR-71 ready and waiting in case the primary aircraft broke, allowing the fliers to change planes. The next level of priority had a second crew in the cockpit of another SR-71, waiting for the word that the primary aircraft had broken and for them to launch. The highest level of insurance had a spare SR-71 flying the same route directly behind the primary aircraft by about thirty minutes. If the primary crew aborted for any reason, the backup crew was airborne and ready to fly the mission. These readiness levels were determined by the priority of the mission and the amount of time available to gather the particular intelligence, labeled the "window." The particular mission window could be anywhere from several hours to possibly only twenty to thirty minutes, depending on the intelligence target and where it was located.

Each Det commander was responsible for getting the mission accomplished

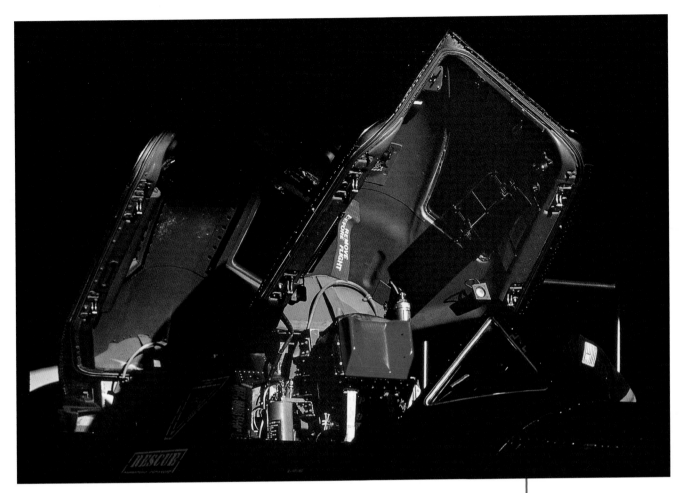

Above: Canopies open. If the pilot ordered a bailout, it was very important for the RSO to eject first from the rear cockpit. If the pilot went first, the pilot's canopy or ejection seat could hit the RSO canopy or ejection seat—a dangerous collision at any speed. *USAF*

Left: The mobile crew in 1987 at Beale AFB has their car pointed directly at the SR-71 inside the hangar about ready to start engines. The white van is the PSD vehicle that brought the crew out to the plane in their pressure suits. *Lockheed Martin*

One of Beale's KC-135Q model tankers refueling the SR-71. The Q-model tankers used 670 gallons of water, injected into the engines, to gain additional thrust on takeoff. Each J-57 engine produced around 10,000 pounds of thrust dry and approximately 13,000 pounds of thrust wet. The water injection lasted about two and a half minutes during takeoff. *Lockheed Martin*

safely. Besides high-quality maintenance troops taking care of the Blackbirds, he had several key personnel at his disposal. For every operational mission, a backup crew was ready to fly in case the primary crew became sick or some other unforeseen reason precluded them from flying. The backup crew, called the "mobile crew," acted as the eyes and ears of the primary crew, anticipating problems that could possibly develop during the mission. By tending to all the smaller tasks, the mobile crew allowed the primary flyers to stay focused on the mission at hand. If the primary crew needed anything at all, the mobile crew was their sole point of contact.

They were named the mobile crew because they were always on the go. Their primary duties were to make sure the plane was 100 percent ready for the flyers and to get the SR–71 into the takeoff position, unimpeded, at the precise takeoff time. The mobile pilot and RSO accomplished a preflight of both cockpits before the primary flyers arrived at the aircraft, ensuring they were set and everything was in perfect order up to the starting engines checklist. There was a tremendous amount of trust between the primary crew and the mobile crew. On several occasions, because of time constraints, crews have literally started both engines immediately after getting into the cockpit. They had complete confidence in the mobile crew's ability to set all the cockpit switches in their correct positions and to check out every single item. The mobile crew had a wealth of classified data, checklists, and worldwide contact information at their disposal ready to assist the flyers. The car was equipped with two UHF radios to communicate with the crews in the plane or controllers in the tower. The check-out for mobile duty came shortly after the seventh SR-71 training flight at Beale.

The term for hazardous objects found on an airport's ramps, taxiways, or runways was FOD (foreign object debris). The mobile crew preceded the SR-71 to the runway in a military vehicle (called the "mobile car") to make sure nothing on the ground could cut the SR-71's tires or hinder its taxi route. The tire footprint of the aircraft was the heaviest in the Air Force inventory, with each tire inflated to 400 psi. Taxiing over a nut or bolt would be like a hot knife going through butter. Perhaps the mobile crews should have been pushing vacuum cleaners . . . now there's a thought!

At each Det there were approximately five to seven KC-135Q tankers and crews from Beale. The tankers and crews were not assigned to the Det commander, but to one of the host tanker units at Mildenhall or Kadena. For commonality of function and maintenance support, it made eminent sense for the tankers to belong to the host tanker unit rather than the Det commander; less for the Det commander to worry about. Each host unit had a designated officer that coordinated all the KC-135Q tanker operations with the Det prior to every mission. If a particular mission called for more tankers, they were flown in from Beale to augment the operation. Occasionally, tankers had to deploy to other locations depending on where the SR-71 was flying and where the air refueling tracks were located. It was not uncommon to have two or three KC-135Qs sent to locations in Spain or Saudi Arabia from Det 4, or to the Philippines or Diego Garcia from Det 1, to support a special SR-71 mission in that part of the world.

The outboard portion of the wing's leading edge has a slight twist to it. This negative camber moves the center of lift inboard to relieve loading on the nacelle structure. It also improves the maximum lift characteristics of the outboard wing and enhances crosswind landing capability. *Lockheed Martin*

CHAPTER 11

Preparing to Fly the Mission

Arriving crews at each Det were not allowed to fly for a minimum of forty-eight hours in order to rest and adjust to the new time zone. During training, every pilot and RSO was screened for usage of specific drugs to help them sleep or stay alert. They could request a limited number of "downers" to aid in sleeping before a mission and "uppers" to help stay awake during long missions. Few Habus ever took the pills; adrenaline always kicked in on difficult missions.

To organize a work schedule among three crews, a "flying ladder" was established. The crew at the top of the ladder was the primary crew for a mission, and on the next rung down was the backup crew that also acted as the mobile crew. The crew at the bottom of the ladder just finished flying and had a few days off to rest and relax. The only thing that changed the rotation of the ladder would be a crew member getting sick or injured. SR-71 crews always flew operational sorties as a formed crew; we did not mix and match crew members as done on other Air Force aircraft. If one crew member was unable to fly then, for all practical purposes, so was the other crew member. (Don and I were convinced vodka tonics with a twist of lime kept colds and flu at bay!)

The day prior to each mission, the primary and backup crews met in Det operations to go over all the details with the mission planners. These were dedicated experts in areas such as SR-71 navigation, enemy defenses, sensor operations, international air traffic control procedures, and other associated operational flying criteria. If there were any questions on the operational aspects of flying the SR-71, the planners were the main source for answers.

During mission planning, the pilot and RSO discussed in detail critical items along the route of flight, such as air refueling track locations, number of tankers, fuel offloads, bank angles at each turn point, Mach numbers, fuel status, and fuel minimums. They further discussed high-threat areas where

Opposite: Crew members have the glamorous job; maintenance personnel do the hard job. They work throughout the night just to prepare an SR-71 for its next flight. If an SR-71 returns with a "Code One" flight, meaning no aircrew discrepancies for the plane, it takes a minimum of twenty-four hours to prepare the aircraft for the next mission. Here the left spike is being worked on. *USAF*

enemy defenses might be encountered, SAM site locations and the closest point of approach (CPA) to unfriendly airspace, various abort scenarios, diversions, alternate bases to land, and weather forecasts. Crews would do a "what if?" drill along the entire route of flight during mission planning while there was time for discussion. Time was a rare luxury flying at Mach 3+, and there was a lot to consider of every mission!

The missions flown out of Okinawa were not as demanding as those from Mildenhall, although both had their challenges. The tight geographical confines in Europe made flying there more difficult. For that reason, new crews flew their first operational missions out of Okinawa before setting off to Mildenhall.

Intelligence requirements of the Vietnam War dominated Det 1's early missions, starting in March of 1968. Flight durations varied from two to well over five hours, depending on the number of targets scheduled on the mission and the number of refuelings. As the war was winding down in 1972, Det 1 began to broaden its area of interest to include North Korea by flying the SR-71 directly down the Korean DMZ, gathering intelligence far to the north. Sorties were also flown off the coast of China and the Soviet Union. Primary targets were the Soviet's largest naval fleet headquarters at Vladivostok and their

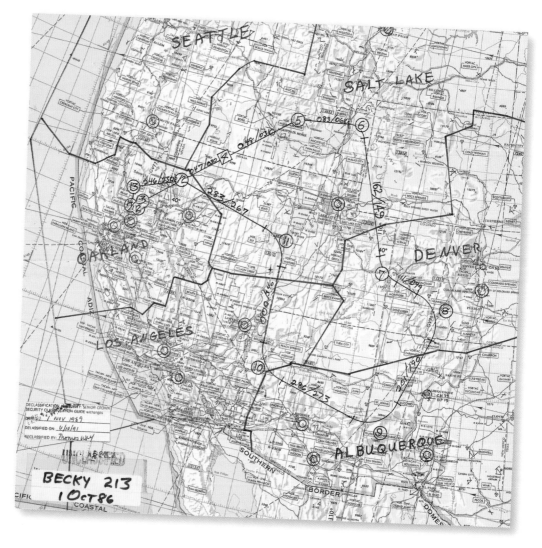

Each destination point (DP) for this training mission called "Becky" is sequentially numbered on the map. The air refueling track is from DP 4 to DP 5, after which the climb and acceleration to 71,000 feet and Mach 3.0 commences. The level off occurs at DP 7. The descent into Beale starts after passing DP 11. Flight planning for an operational mission is considerably more complex. *USAF*

```
* * * *
* * * * * * * *        UNCLASSIFIE[
        PILOT MISSION SUMMARY              PAGE
  BECKY SIM   213  30/04/90                  4
  DIST                              TACAN
  LEG   BACK  ETE  TH  ALT CALL   CH
  PT AR/DEST SET  TET  MH FUEL RAD  RGE

        MACH 3.00

DP  8     262   TSA 00.5 202 71.0 HBU   96
  TSP   1253    73 124.2 191 46.8 259    29

BANK ANGLE = 32   TEMP T + 3    TAS 1724

DP  9     281   TSA 09.1 288 72.2 ONM  115
  TSP    991    71 133.3 276 40.5 277    54

BANK ANGLE = 32   TEMP T + 3    TAS 1727

DP 10     257   TSA 09.8 000 73.6 DRK    88
  TSP    710    57 143.0 346 33.9 254    24

BANK ANGLE = 32   TEMP T + 3    TAS 1731

DP 11     299   TSA 08.9 285 74.8 ILC   110
  TSP    453    57 151.9 270 28.1 031    29

BANK ANGLE = 32   TEMP T + 3    TAS 1735

          147       05.4 285 74.7 BAM    59
  S/D     301   152 157.3 270 27.1 124    47

           56       03.8 283 60.0 NFL    82
  IKP     210    61 201.1 267 26.7 005    61

           20       02.2 283 45.0 HZN    88
  IKP     174    25 203.3 267 26.4 345    59

* * * *
* * * * * * * * * * * * * *     UNCLASSIF[
```

Above: Pilot Duane Noll (left) and RSO Tom Veltri eating their required high protein/low residue meal at the dining facility in PSD at Beale. This meal ensures crews are fully nourished prior to every flight (a few crew members have had disastrous results inside the pressure suit from whatever they ate the night before). *Paul Crickmore*

Left: One page of the computed flight plan for "Becky." Each DP has various information listed. Back set is the distance the aircraft will begin each turn at a specific bank angle. ETE and TET are Estimated Time Enroute between each DP and Total Estimated Time from takeoff. TH is True Heading and MH is Magnetic Heading at each DP. Alt and fuel is the approximate altitude and fuel quantity at each DP. *Rich Graham*

nuclear submarine facility at Petropavlovsk, located on the southern end of the Kamchatka Peninsula. Those sorties ranged between two and four hours. Between late 1987 and early 1988, four long-endurance missions were flown into the Arabian Gulf, two of them lasting well over eleven hours. A long time to be strapped into an ejection seat, unable to move with a pressure suit on!

Missions out of Det 4 gathered intelligence on three basic areas of interest. The Barents Sea missions were flown primarily to gather intelligence on the Soviet Union's second largest naval facility near Murmansk. In the Baltic Sea, the SR-71 was able to glean intelligence on the then-Warsaw Pact countries of Poland, Lithuania, Latvia, and Estonia. Flying in West Germany, the SR-71 gathered intelligence on East Germany, Poland, and the Czech Republic (then Czechoslovakia). Crews flew any combination of the three areas, sometimes all three on a single mission. Long-endurance missions into the Arabian Gulf region were flown out of Det 4, as were post-strike reconnaissance missions over Libya in 1986.

After mission planning was finished to everyone's satisfaction, both crews stored their classified materials back in the safe and were free to do as they pleased until crew rest time came around. Most crews enjoyed relaxing the day before their mission. Okinawa was a great place for getting in a round of golf, playing tennis, driving to one of many pristine beaches, or shopping around the island. For the crews in England, shopping, tennis, touring, and taking in the extensive history of the country were favorite pastimes. At both Dets, each crew had a civilian car to drive around.

With a wake-up time around three hours before takeoff, Habus had to get to bed much earlier than normal Air Force pilots. For a 0800 takeoff, the following schedule was typical.

2100: Sleep
0500: Wake up, shower, dress
0530: Preflight meal
0615: Mission briefing start
0645: Physical exam and suit up
0715: Arrive at aircraft
0730: Start engines
0800: Takeoff

Crew activities leading up to the flight were basically the same at each Det. After meeting in the hallway or in one of the bachelor officer's quarters (BOQ) rooms, the primary and mobile crew drove to a dining facility that served their mandatory preflight meal. The meal consisted of steak and eggs with coffee, milk, or apple juice. That was considered a high protein/low residue meal, supposedly ideal for being trapped inside a pressure suit for the next three to five hours or longer. On twelve missions that meant well over eleven hours of flying! The preflight meal was also a good opportunity to discuss in private any particular needs or concerns that might have developed since the previous day's mission planning. Mission planners were already at Det operations, busily preparing for the briefing. Maintenance preparations on the aircraft began long before the crews

woke up—maintenance personnel were the real early birds.

After eating, both crews drove to the Det for the mission briefing. Once inside operations, the primary crew retrieved their classified checklist, route maps, and flight plans from their safes. They took all their mission materials into the briefing room to look over any last-minute details before the start of the briefing. The primary purpose of this was to appraise the Det commander of the mission being flown and to make sure everyone in the briefing room was "singing from the same sheet of music." It was extremely important for everyone attending the briefing to know precisely what the mission entailed, because operational sorties were conducted in radio silence to the maximum extent possible. Personnel required to attend the mission briefing were the Det commander, operations officer, primary and backup crews, mission planners, weather briefer, maintenance officer, and crew chief for the plane (unless he was taking care of a problem with the plane).

The SR-71 program always flew under the operational concept that no news was good news. As long as everything was

Duane (right) and Tom get their required preflight physical. Some crew members had the "white coat" syndrome, where blood pressure is out of limits as soon as the cuff is wrapped around their arm. For all practical purposes that would ground both crew members from flying and the mobile crew (back up) would then fly the mission. *Paul Crickmore*

Terry Pappas (left) and John Manzi suited up. The tubes connected to their helmets are used for pressure testing the suit. Notice the stirrup lying on the floor. PSD would attach this to the bottom of each boot with Velcro straps. After the stirrups were attached, boots zipped, and outer layer zipped up, the crew was ready to go. *Lockheed Martin*

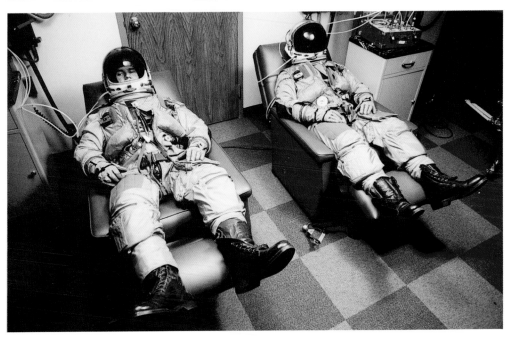

One of the PSD vans used to transport the crew out to the plane. The reclining chairs are bolted to the floor and cooling air is provided from the portable liquid oxygen converters through the hoses connected to the pressure suit. *USAF*

proceeding normally with the mission, there was no need to communicate. This provided an extra measure of communication security. For many years, standard Air Force radio calls were made with air traffic control on all operational missions. Around 1978, the National Security Agency (NSA) came down from Japan to give Det 1 a security briefing. Posing as fishing trawlers just off the coast of Okinawa, the Soviets were easily picking up our radio transmissions and passing the information along to other communist countries. Following NSA's recommendations, radio-silent procedures were developed for all operational missions worldwide. It took several months to develop these procedures, but eventually crews were able to takeoff, fly, and land without talking to anyone. Crews were allowed to talk over the radios in the interest of safety or getting the mission accomplished.

At the completion of the briefing, the Det commander generally added any pertinent comments. Most commanders reinforced the fact that there was no mission worth losing a plane for. Pilots had to fly safely and not

Doug Soifer exits the PSD van followed by his pilot, Mike Smith. They were flying an operational mission out of Okinawa in February 1988. *Doug Soifer*

take any unnecessary chances. The mission could always be flown another day. Crews appreciated those words, because it took some of the pressure off their shoulders. No Det commander would ever second guess a crew's reason for aborting a mission.

After the briefing, everyone departed to their own area of expertise. The tanker commander checked on the status of the KC-135Qs that should have been airborne

Bill Orcutt talking over the status of the plane to Terry Pappas (left) and John Manzi before they head up the stairs. The two hoses on the back of each helmet connect to the aircraft's oxygen system, providing 100 percent oxygen to prevent the crew from getting the bends at high altitudes. *Lockheed Martin*

Opposite bottom: Terry and John stepping into their respective cockpits, placing their right foot on the ejection seat. Once stepping onto the seat, they place each foot on the floor of the cockpit while holding onto something solid. The first order of business is to get the stirrups attached to their retraction cables located on the ejection seat bottom. Upon ejection, the retraction cables lock their feet firmly in place, mandatory for a Mach 3 ejection. *Lockheed Martin*

by now. The mobile crew gathered up all their classified materials and the mobile kit bag and departed for the hangar to begin the preflight of both cockpits. The mission monitor gathered up his materials to follow the aircraft's route of flight based on timing and departed to a highly secure location on base. If the SR-71 crew had problems while airborne, the mission monitor was the first to be notified. He could then inform others in the chain of command via secure means.

The operation officer, usually a former crew member, also acted as the tower officer. His primary job in the airfield tower was to ensure there were no delays in taxi, takeoff, or landing for the SR-71. Tower personnel did not have Senior Crown clearances and, therefore, no knowledge of the aircraft's capabilities or, in the event of an emergency, what was needed.

The mobile crew drove to the hangar to begin their preflight check of the SR-71 assigned to the mission. They took time to chat with the crew chief, other maintenance personnel, and the tech reps working on and around the plane, getting it ready for flight. In each cockpit, the pilot and RSO mobile crew went through their respective preflight checklists, stopping at the starting engines checklist. If the mobile crew found anything wrong with the aircraft, the crew chief was the first to be notified.

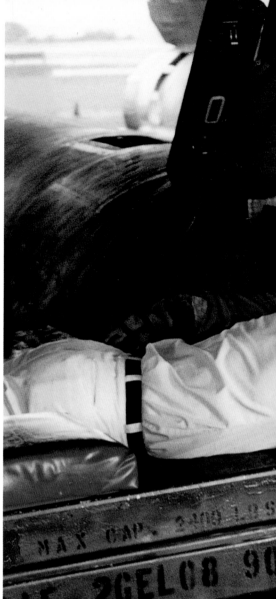

After the briefing, the primary crew departed to the PSD facility to begin preparing for the pressure suit. Changing out of their flight suits into cotton long john underwear, each crew member began the physical examination required before every flight. The examiner recorded their blood pressure, pulse, weight, temperature, sinus problems, time to bed, time awoken, and time and contents of last night's meal.

Prior to every mission, PSD technicians checked over each crew member's suit and laid it out in a precise manner to expedite putting it on. If the crew member and PSD technician were experienced, it took around

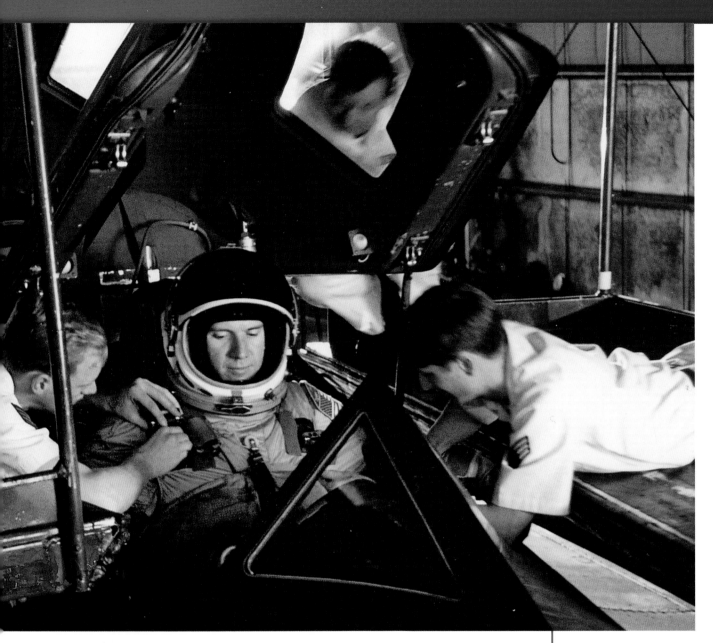

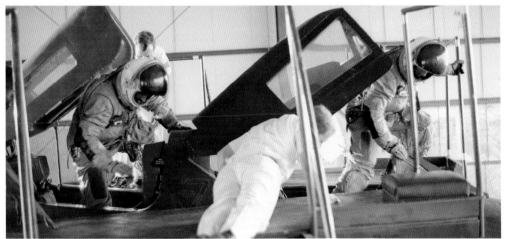

Above: Once seated inside the cockpit, it was up to PDS technicians to make all the connections. The oxygen supply hoses, electrical face heat, stirrups, communications cord, cooling air supply, seat kit attachments, lap belt, shoulder harness, and parachute harness all had to be connected securely and safely. *Lockheed Martin*

Above: The pilot's view of the cockpit. The yellow T-handle near the right windshield was called the "dinghy stabber." The Air Force had experienced inadvertent life raft inflations inflight. Several were fatal because it forced the pilot into the control stick. The dinghy stabber had a sharp point for the pilot to stab and deflate the life raft. *Tony Landis*

Right: The massive rudder pedals adjust fore and aft for various pilot heights. Aircraft braking is applied by pushing on the top of each rudder pedal. Nose wheel steering is applied by movement of the rudder pedals. The yellow "T" handle left of the control stick is for emergency gear extension. *Rick Poole Photography*

fifteen minutes to get the suit, boots, gloves, and helmet on. Then the suit had to be checked out thoroughly before leaving the PSD facility. With the crew member sitting in an overstuffed reclining chair, the suit was connected to a pressure testing unit and fully inflated to make sure it held pressure satisfactorily and that the communications and face heat worked properly. Testing results were recorded for trend analysis of each suit.

Once the suit checked out, the PSD technicians made final adjustments to its exterior. When both crew members were ready, they stood up and walked out to the waiting PSD van, ready for the drive to the aircraft. If there had been any problems with the SR-71, tankers, or weather, the mobile crew would have let the flyers know prior to departing PSD. For the ride out to the plane, the crews sat in recliner chairs bolted to the floor of the van. After a short drive to the hangar, the van parked directly in front of the waiting SR-71.

With about forty-five minutes to go before takeoff, the PSD technicians exited the van and headed up the portable steps to the cockpits. The mobile crew went inside the van to discuss the status of the plane, passing along any last-minute information concerning the mission. If the crew chief needed to discuss anything

with the flyers, he also joined them in the van. Before anyone was allowed to go up the stairs to the cockpits, they had to rid themselves of all personal items, including watches, rings, coins, pens, pencils, badges, or anything else that could possibly be dropped inside the cockpit. Lying flat on their stomachs at the top of the ladder's platform, PSD technicians reached inside each cockpit and moved hoses and connections, preparing to attach the crew member's pressure suit to the ejection seat.

Following a thumbs-up signal from PSD at the top of the ladder, the flyers exited the rear of the van and walked to the bottom of the steps. They climbed up the stairs, bent over to clear the open canopy, slowly made their way to a standing position on the cockpit floor, and then sat down in the ejection seat. Methodically, the PSD technicians reached down each side and made all the necessary connections to mate the pilot and RSO to the ejection seat. During training, crews were instructed to put their arms out each side of the cockpit and not help PSD make the connections; in a pressure suit it would be extremely difficult to make many of the connections and to be able to tell if something was improperly connected. A crew member's survival was totally dependent on PSD to make all the right connections securely.

CHAPTER 12

Flying the Mission

After PSD was finished, each crew member began his own individual interior cockpit checklist. The crew chief put his headset on and checked in with the pilot and RSO. A typical reply from the crew chief was, "Everything's good down here, ready to start engines whenever you are." The pilot replied with an estimated time. This is when most crew idly chatted with the crew chief. Once the pilot was ready to start engine, the atmosphere became professional, and the checklist script was cast in stone. When the mobile crew observed the pilot getting close to engine start, they returned to the mobile car and monitored the UHF radios in case they were needed by the flyers. They positioned the car directly in front of the SR-71 and patiently waited. By observing the crew chief's action around the plane, the mobile crew knew precisely where they were in the checklist.

Decompression sickness, or the "bends," is typically associated with scuba divers when they rise to the surface of the water too rapidly. It's caused by the formation of nitrogen gas bubbles in body tissues, primarily around joints, and can be extremely painful, sometimes to the point of being debilitating. Bubbles in the blood stream can even cause strokes or death. The same thing can happen to crew members if they fly at extremely high altitudes. To preclude getting the bends, SR-71 crew members pre-breathed 100 percent oxygen just prior to starting engines, once the faceplate was lowered on the helmet. After lowering and locking the face plate, the pilot turned on the exterior, red beacon light, alerting everyone in the hangar engine start was imminent, and to put on their ear protectors.

Opposite: Operating on a test stand, the J-58's afterburner section glows red-hot. Shock diamonds in the exhaust occur when gas exiting a nozzle at supersonic speeds is at a different pressure than the outside atmosphere. *Pratt & Whitney*

Right: Inside the cockpit, Terry begins his pilot's checklist. You can see the bailer bar in the up (vertical) position and his face exposed to outside air. Prior to starting engines, he and the RSO will rotate the bailer bar downward, locking it into place, starting the flow of 100 percent oxygen. *Lockheed Martin*

Top: Because of the high flash point of the JP-7 fuel, a unique liquid ignition system was developed. The liquid was triethylborane (TEB), a chemical that burns spontaneously when exposed to the air (as seen in photo), which in turn ignites the JP-7 fuel. The tank held enough TEB for sixteen "shots" to ignite both the main engine and afterburner section. *Don Emmons*

Bottom left: This starter adaptor eventually replaced the Buicks in the early 1980s. The geared shaft on top connected directly to the J-58 engine. Lifted up to the engine by the yellow handholds and locked into place with the lever in the front, air hoses from several ground starting units on the massive compressed air tanks at Beale were connected to the inlets to provide engine rotation. *Rich Graham*

Bottom right: The Buick starting cart. On the top front (beneath the slightly open protective cover) is a mechanical geared shaft that connects the Buick to the bottom of the J-58 engine. The throttle for both V-8 engines is on the bottom center of the control panel with a knob on top. There are no mufflers installed and the sound inside the hangars is unbelievable. *Mike Hull*

Starting the massive J58 engine was accomplished by a direct mechanical drive shaft to initiate engine rotation. The large starting cart used to turn the engine over was called a "Buick" because it originally had two large-block Buick V-8 engines, mounted beside each other, providing over 600 horsepower. Through a series of reduction gears, the two engines powered a vertical shaft, extending upward and connecting directly to the bottom of the J58 engine. No air starting carts were capable of supplying a sufficient volume of air to rotate the massive J58 engine until the early 1980s, when air start capability was finally developed.

After both engines were started, a series of operational checks of the flight control, generators, fuel system, air refueling, radios and navigation systems was carried out. When all the checks were satisfactory, the crew chief walked up the portable ladders and lowered each cockpit canopy, holding it down while the pilot and RSO locked

Left: The crew chief signals for the pilot to begin a left turn out of the hangar and onto the ramp area where he will then follow the yellow line. At idle power it was very easy for the SR-71 to accelerate faster than needed during the taxi out to the runway. However, the pilot didn't want to ride the brakes as they could overheat easily. *Lockheed Martin*

Below: Taxi out with the taxi light on. The mid-span location of the engines minimizes drag and interference effects of the fuselage at high Mach speeds. The noticeable inboard cant and droop of the nacelles gives maximum pressure recovery inside the inlet for high altitude Mach 3+ cruise. *Tony Landis*

Right: The takeoff and landing card is filled out prior to takeoff by the crew members. Notice the numbers if they have to land immediately after takeoff. The assumption is they will burn 4,000 pounds of fuel before landing, leaving a gross weight of 116,000 pounds. The final approach speed is 221 knots or 254 mph! *Rich Graham*

Far right: Last chance inspection on Okinawa. While the crew chief remains in direct view of the pilot, his assistants methodically check over the SR-71 for any abnormalities underneath the plane. If something didn't check out, the crew chief could plug his headset into the interphone panel and talk directly to the pilot or RSO to discuss the problem. *Tony Landis*

Bottom: Aircraft 960 and 967 starting their engine run-up on Okinawa with the Pacific Ocean in the background. It's very rare to see two SR-71s in the run-up area at the same time. Most likely, the mission was high priority and one of the aircraft is either an airborne or ground spare in case the other one has a malfunction. *USAF*

each canopy securely in place. The crew chief and all his assistants busied themselves removing the electrical, air conditioning, and any other connection to the SR-71. The plane was now on its own internal systems. The crew chief professionally walked about fifty feet in front of the plane, did a military "about face," and came to "parade rest," waiting for the pilot to let him know it was time to taxi out of the hangar.

A quick flash of the taxi light, and the crew chief had the chocks removed by his assistants. The pilot advanced the throttles forward slightly to start the taxi process. The mobile crew preceded the SR-71 out to the runway, looking for any debris on the taxiways. Following behind the SR-71 came a small caravan of maintenance vehicles to make sure the aircraft passed its final inspections prior to takeoff.

Just short of the runway, the pilot taxied the plane to an open area to perform a run up of each engine individually to military power—100 percent (RPM). Once in position, chocks were again put in place, but to be safe the pilot still held the brakes tightly, while advancing the left throttle to military power for a minimum of thirty seconds. Onboard aircraft recorders logged all engine parameters that would be used later by maintenance to establish a baseline and troubleshoot engine problems. This "last chance inspection" was to ensure everything was working satisfactorily.

The pilot and RSO reviewed the takeoff and landing data card to check the numbers for the takeoff acceleration check speed, refusal speed, rotation speed, takeoff speed, and single-engine speed. The acceleration check speed was the minimum speed the aircraft should have at a given distance down the runway. If the speed was less than the check speed, something was wrong, and the pilot had to quickly decide to continue the takeoff or abort on the runway. The refusal speed was the maximum speed the aircraft could accelerate to and still stop on the remaining runway. At rotation speed, the pilot began to initiate back pressure on the

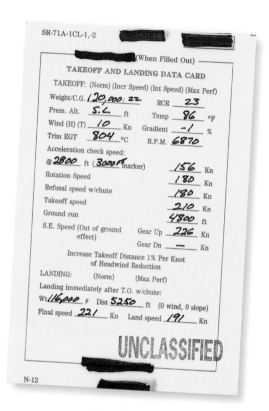

SR-71A-1CL-1,-2

(When Filled Out)

TAKEOFF AND LANDING DATA CARD

TAKEOFF: (Norm) (Incr Speed) (Int Speed) (Max Perf)

Weight/C.G. 120,000 / 22		RCR 23	
Press. Alt. S.L. ft		Temp 86 °F	
Wind (H) (T) 10 Kn		Gradient -1 %	
Trim EGT 804 °C		R.P.M. 6870	

Acceleration check speed:

@ 2800 ft (3000 ft marker) _156_ Kn

Rotation Speed _180_ Kn

Refusal speed w/chute _180_ Kn

Takeoff speed _210_ Kn

Ground run _4800_ ft

S.E. Speed (Out of ground effect) Gear Up _226_ Kn

Gear Dn _—_ Kn

Increase Takeoff Distance 1% Per Knot of Headwind Reduction

LANDING: (Norm) (Max Perf)

Landing immediately after T.O. w/chute:

Wt _116,000_ # Dist _5250_ ft (0 wind, 0 slope)

Final speed _221_ Kn Land speed _191_ Kn

UNCLASSIFIED

N-12

SR-71B model trainer aircraft 956 at the run-up area on Beale AFB. The aircraft has ventral fins under each nacelle to aid in lateral stability because of the raised rear cockpit. The crew chief is checking underneath for anything unusual. One assistant crew chief at the far right holds the sledge hammer to remove the chocks. *Lockheed Martin*

SR-71 968 holding short of runway 23R at Det 1 on Okinawa. Just on the other side of the SR-71 is the maintenance vehicle that followed the aircraft while taxiing. It was equipped with a platform and stair arrangement that could pull up to the SR-71, allowing crew members to exit in case of a ground emergency. *USAF*

stick to slowly lift the nose wheel off the runway. At takeoff speed, the SR-71 should be airborne. If an engine was to flame out on takeoff, the single-engine airspeed was the minimum airspeed to fly. After all the before-takeoff checks were accomplished, the pilot gave the crew chief the "remove chocks" visual hand signal. Quite often, the massive wooden chocks became lodged into the concrete from the forward thrust of the SR-71's engines during run up. The world's fastest and highest flying jet plane had to be released to fly with a sledgehammer by the assistant crew chiefs banging away to dislodge the chocks. After the chocks were removed, the crew chief and pilot exchanged salutes.

Using binoculars, the tower officer knew precisely how things were progressing at the end of the runway and politely asked tower personnel to give the waiting mobile car and SR-71 pilot a flashing green light. With about a minute to go, that signaled the mobile crew to drive onto the runway and gave clearance for the SR-71 to take off

when ready. The mobile crew proceeded onto the runway to check for any objects lying in the way.

The SR-71 had a unique departure end barrier (designated BAK-11/12 barrier) developed to stop the aircraft on the runway if all other means failed. It consisted of a set of pressure-sensitive switch mats located about six feet apart, built into the runway to sense the nose tires rolling over each of them. Timing computers quickly calculated the speed of the SR-71 and fired off an air pressure charge that quickly pulled the barrier cable up into the air in time to snag the main gear and stop the aircraft. Although there would be minor damage to the landing gear, it was better than losing an aircraft to a crash or fire. A barrier was located at Beale on runway 14 and at Kadena on runway 5R, but there were none at Mildenhall. The barriers were armed and disarmed in the tower. The air charging process took about five minutes before the barrier could be armed. Once sufficient air pressure was available, a green "charged" light illuminated in the tower. On landing, the tower controller informed the pilot that the "barrier is armed." The tower officer was there to ensure the arming and disarming process was successful. There were very few totally successful barrier engagements by an SR-71. Consequently, crews never planned on using it to stop them.

The pilot added power and taxied onto the runway and stopped straight ahead on the centerline. Once the mobile car cleared the runway, the pilot was ready for takeoff. The pilot engaged nose wheel steering and waited for the exact takeoff time. The RSO had a clock in his cockpit accurate to one-hundredth of a second and initiated the takeoff countdown by saying, "five . . . four . . . three . . . two . . . one . . . hack." Accounting for the lag time in the throttles to spool up the engines, the pilot began advancing them to full military power about the time the RSO said "three" and released the brakes on the "hack."

On 10 October 1968, pilot Abe Kardong and RSO Jim Kogler were taking off from Beale AFB when a tire blew out, sending shrapnel up into the fuel tanks, creating a massive fire. Abe tried to abort the takeoff. Jim ejected safely while Abe continued to steer to the center of the BAK-11/12 barrier at the far end of the runway. The barrier engagement was unsuccessful, but the mobile crew got Abe out safely. *USAF*

MANY PEOPLE BELIEVE we refueled after takeoff because the aircraft leaked fuel so profusely that we needed to fuel up quickly. We had to refuel right after takeoff for only one reason, and it wasn't because we leaked JP-7 fuel on the ground. Yes, the plane does leak fuel, but not enough to require refueling after takeoff.

The JP-7 fuel reaches temperatures well over 300 degrees F. during Mach 3 cruise, making the fumes in each of the six fuel tanks very volatile and potentially explosive. The metal skin of the aircraft approaches 400 degrees F., adding to the volatility of the fuel inside the tanks. One of our aircraft limitations was a maximum speed of Mach 2.6 without an inert atmosphere inside the fuel tanks.

The aircraft had three liquid nitrogen Dewar flasks containing 260 liters of liquid nitrogen, located in the nose wheel well. The only way to ensure 100 percent inert atmosphere in each fuel tank was to refuel the plane inflight completely full of JP-7, allowing ambient air in each fuel tank to vent overboard. Once full of fuel, gaseous nitrogen would now dominate each fuel tank's empty space above as it burned off JP-7. The nitrogen gas pressurized each fuel tank to 1.5 psi above ambient pressure and inerts the space above the heated fuel to prevent autogenous ignition. This is why we refueled after takeoff. Then we could safely accelerate beyond Mach 2.6.

There was one other way of achieving tank inerting, called a "Yo-Yo," but this was a maintenance nightmare. A few of our missions required the SR-71 to accelerate to Mach 3+ right after takeoff with a 65,000-pound fuel load. The Yo-Yo procedure had the crew chief completely refuel the plane to full tanks of 80,000 pounds of fuel. Then, with the nitrogen pressurization system working, they de-fueled 15,000 pounds of JP-7, ending up with a 65,000 pound fuel load and a plane that was capable of going immediately to Mach 3+.

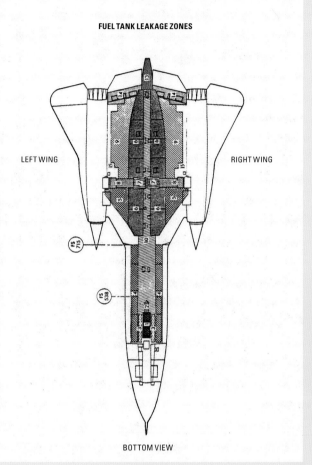

FUEL TANK LEAKAGE ZONES

LEFT WING

RIGHT WING

BOTTOM VIEW

Precise reading of the engine instruments was time-consuming and, more importantly, there were other things needing watching during takeoff. Most pilots memorized where the needles should point on each gauge. Once satisfied by the needle positions of the rpm, exhaust gas temperature (EGT), nozzle position, and oil pressure gauges, the pilot lifted up both throttles and advanced them full forward into maximum afterburner (AB). In full AB, the pilot made another quick scan of the engine instrument readings.

The goal now was to keep the SR-71 heading directly down the centerline of the runway while slowly moving the stick back at the rotation speed of 180 knots (205 mph). With a smooth, continuous aft movement of the stick, it took about five seconds from the start of rotation until liftoff. The SR-71 broke ground at 210 knots (240 mph). Once airborne, the pilot continued to rotate the nose upward while raising the landing gear lever to the "up" position. As the aircraft accelerated through 0.5 Mach, the surface limiter caution light came on, telling the pilot to rotate and engage the surface limiter T-handle. Once engaged, the surface limiter restricted roll and rudder movements at higher airspeeds to preclude over stressing the aircraft. The

TABLE OF FUEL TANK LEAKAGE LIMITS

Zone	Zone Description	Max. Leakage Allowable
1	NLG Wheel Well	10 Drops/Minute/Side
2	NLG Trailing Door Compartment	30 Drops/Minute
3	E-Bay and R-Bay	25 Drops/Minute/Bay
4	Left and Right, Forward and Aft, Mission Equipment Bays	25 Drops/Minute/Bay
5	Fwd Lower Wing, Including Fwd Wall of MLG Wheel Well	150cc/Minute/Side
6	Fuselage Fillet Area	Leakage unimportant, but no single leak exceeding 50 Drops/Minute/Side
7	Inboard of MLG Wheel Well (Excluding Fwd Wall of MLG Wheel Well)	5 Drops/Minute/Side Leaks not permitted in fuel plumbing.
8	Outboard MLG Wheel Well (Excluding Fwd Wall of MLG Wheel Well)	Dampness, no dripping permitted. Leaks not permitted in fuel plumbing.
9	Aft, Lower Wing	60 Drops/Minute/Side
10	Fuselage	Leakage unimportant, but no single leak exceeding 50 Drops/Minute
11	Top, Wing	Leakage unimportant, but no single leak exceeding 60 Drops/Minute/Side
12	Drag Chute Compartment	Dampness, puddling not permitted
13	Tailcone (excluding vent and dump line leakage)	10 Drops/Minute
14	Aft of Rear Beam	Leakage unimportant, but no single leak exceeding 60 Drops/Minute/Side
15	Dry Bay Area	950cc/Minute/Side

Left: Maximum allowable fuel leakages, according to the SR-71 maintenance handbook. Leaking JP-7 fuel was collected in a large shallow drip pan placed underneath the aircraft in an attempt to keep it off of hangar floor. *USAF*

The Pratt & Whitney J-58 engine is designed for continuous afterburner operations up to a compressor inlet temperature of 427 degrees C (800 degrees F). Between Mach 1.8 and 2.0, the large six compressor bleed tubes (three per side) take air from the fourth stage of the compressor and route it around the turbine to provide additional thrust and cooling at the exhaust. *Pratt & Whitney*

pilot maintained 400 knots until reaching 0.9 Mach, then held the speed constant until reaching his cruise altitude. The climb pitch attitude was initially around thirty-five to forty degrees depending on the fuel load. All subsonic cruising was at 0.9 Mach. The departure was also flown radio silent; we spoke to no one. The monitoring radar control facility identified the SR-71 by the identification friend or foe (IFF) code appearing on their radar screen. Most operational sorties started off with a refueling after takeoff.

CHAPTER 13

Refueling

No story on the SR-71 would be complete without an understanding and appreciation of just how valuable the KC-135Q model tankers and their crews were to the successful and safe completion of every sortie. The SR-71 would be a very short-ranged aircraft were it not for air refueling, limited to around 2,000 NM. Multiple air refuelings extended the range of the aircraft to the limits of crew endurance. Many missions have exceeded 12,000 NM. Forward basing of the SR-71 and KC-135Q tankers permitted faster response, shorter range, shorter duration missions, fewer air refuelings, and greater overall efficiency.

The KC-135Q model tankers were unique within the Air Force and given the "Q" model designation. Beale had 35 KC-135Q model tankers flown by the 349th and 350th Air Refueling Squadrons to support the SR-71's global operation. At one time both squadrons were under the 100th Air Refueling Wing on Beale, but in 1976 someone made a very wise decision to place the two air refueling squadrons under the command of the 9th SRW, since they were the exclusive means of refueling and went TDY everywhere the SR-71 went. By the mid-1980s the SR-71 was also occasionally using Air Force KC-10s for refueling.

We also utilized Beale tankers as the normal means of transportation back and forth to Kadena and Mildenhall. Leaving Beale the tanker was called the "deployer"; returning it was the "redeployer." Every two weeks at Beale a

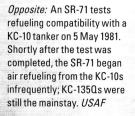

Opposite: An SR-71 tests refueling compatibility with a KC-10 tanker on 5 May 1981. Shortly after the test was completed, the SR-71 began air refueling from the KC-10s infrequently; KC-135Qs were still the mainstay. *USAF*

Right: This is the best sight in the world when you're low on JP-7. As the SR-71 moves into the contact position to receive his fuel, the air flowing over the SR-71 creates a "bow wave" that physically raises the tail of the KC-135, requiring the pilots to re-trim the plane slightly. *Lockheed Martin*

Another test of air refueling compatibility, this time with a Boeing 747. It proved feasible, but the tanker version of the 747 never materialized. *Lockheed Martin*

deployer would transport SR-71 crews, maintenance personnel, and equipment to both Okinawa and Mildenhall, and upon landing, the redeployer would take personnel and equipment back to Beale.

Beale tanker crews worked, lived, and partied together with the SR-71 crews at both Dets. A special bond developed between the tanker crews and SR-71 crews that didn't exist throughout the Air Force. They took considerable pride in their work because of the exclusive SR-71 refueling mission. They also knew that the SR-71's mission success was directly related to inflight refueling. They were always there, in the murk and dark of night, with a full load of JP-7 fuel awaiting a thirsty SR-71.

For a Habus, it was comforting to have Beale tanker crews, who knew you well, doing the refuelings. Throughout twenty-two years of refueling the SR-71, there has never been a missed air refueling because

the tankers were not there. In fact, in several instances, the tanker crews have flown out of their orbits to save an SR-71 with a dire emergency or extremely low on fuel.

KC-135 crews and their aircraft were unique from the rest of the Air Force in several ways. Their aircrews were the only ones certified in our specific radio-silent rendezvous procedures, and their boom operators were the only ones qualified to refuel the SR-71. The Q-model tankers had special plumbing between their fuel tanks, allowing them to transfer JP-4 and JP-7 fuel between various tanks. Their engines could burn either JP-4 or JP-7 fuel. If the SR-71 landed somewhere JP-7 fuel was not available, the Q-model tankers flew in with the fuel and, through the use of transfer hoses on the ground, were able to refuel the SR-71. One of the best advantages of flying the Q-model tankers is that their crews did not have to be on twenty-four-hour

Beale's T-38s and KC-135Q tanker. Eventually, under an Air Force program, the T-38s were flown by younger Beale tanker pilots to help maintain their proficiency. *USAF*

alert status like the rest of the SAC's tankers' crew members.

Another important feature of the KC-135Q tankers was the ARC-50 radio onboard the plane. Both the SR-71 and KC-135Q had this installed. It used standard UHF to communicate, but had a unique feature associated with it that allowed coded communications and variable power levels for the transmitter, up to 100 watts. With the tanker and SR-71 on the same frequency, and the correct codes set in, range and bearing were displayed to each other's aircraft. Over the open ocean, it was not uncommon to pick up the tanker at ranges of three hundred miles or greater.

When more than one tanker flew in formation, they were referred to as a "cell." Depending on the amount of JP-7 to offload, there would be either two, or sometimes three, KC-135Qs in a cell. Often, we took fuel from two tankers (called a "split

offload") and had a third tanker there in case one of them developed transfer problems or had to return because of an inflight emergency. Having three tankers in the cell provided redundancy, if needed.

It was important for the tankers to be in the air refueling track thirty minutes prior to the SR-71's arrival to check weather conditions. The SR-71 had no weather radar capability whatsoever, and, therefore, the crews relied heavily on the tanker's weather recommendations. Frequently, the tankers had to move the air refueling control point (ARCP—the common rendezvous point for the SR-71 and the tankers) because of bad weather, necessitating a different arrival route for the inbound SR-71. Rendezvous with the tankers were always one of two types. A "hot" rendezvous originated from a high Mach number cruise leg and supersonic descent, while a "cold" rendezvous was a subsonic arrival, generally soon after takeoff.

Every time the SR-71 flew, there was a minimum amount of fuel required to have onboard at each ARCP, called "Bingo" fuel. This was computed so that you could arrive at the ARCP, not refuel for one reason or another, and still be able to divert to an alternate base with a minimum of 10,000 pounds of fuel remaining. In theory, whenever crews computed they would be below Bingo fuel at the ARCP, they were

Staying on the boom requires good formation flying skills. The refueling boom has vertical, horizontal, and fore and aft limits. Once connected and locked into the SR-71's air refueling receptacle, an automatic disconnect occurs if the pilot exceeds any of the travel limits. Bouncing around in rough air sometimes made it difficult to stay within the boom limits. *Lockheed Martin*

supposed to abort the mission and proceed to their alternate landing base. In reality, most Habus took an educated gamble rather than abort. As long as the refueling and alternate weather were good, it was a gamble worth taking. Another option was to call the tankers prior to the supersonic descent and have them move the ARCP closer to the SR-71's descent track, saving precious fuel. However, this option was not always available.

The SR-71's refueling system received fuel at approximately 6,000 pounds per minute. All of the six main fuel tanks could be filled up in about fifteen minutes at a refueling pressure of 65–70 pounds per square inch (psi). Once connected and transferring fuel, the boom would automatically disconnect whenever 70 psi of backpressure was sensed in the refueling boom.

Both the tanker and SR-71 crews had the same information on their classified air refueling cards showing: call signs, track locations, refueling altitudes, on load fuel, bingo fuel, and the necessary code used to link up each other's aircraft in order to get bearing and range to each other. Typical refueling altitudes would be between 24,000 to 28,000 feet.

The actual rendezvous was left up to the lead tanker crew to accomplish. The SR-71 crew merely flew on course, straight towards the ARCP at 0.9 Mach, 2,000 feet below the actual refueling altitude. Prior to the SR-71's arrival, the lead tanker navigator had his aircraft commander flying a race track orbit so he could determine the wind drift and timing of their orbits. When the head-on distance to the inbound SR-71 was around ten NM, the navigator had the aircraft commander begin the final 180-degree left turn, planning to roll out about three to four miles directly in front of the SR-71.

Once the SR-71 approached one mile behind, the lead tanker initiated a descent and accelerate maneuver. The maneuver called for the tanker cell to add power, descend 1,000 feet and accelerate to 325 knots. The SR-71 remained at their altitude until sighting the lead tanker visually, then slowly moved in and up to the pre-contact position. At this point the SR-71 crew accomplished their air refueling checklist prior to hooking up on the tanker's boom. Flying in the correct refueling position required a tremendous amount of concentration outside the cockpit, making it impossible for the pilot to check on anything inside the cockpit.

Every SR-71 pilot lowered his seat as he approached the precontact position. The upward visibility to watch the tanker's director lights was very uncomfortable in the normal flying position. To decrease the angle looking upward and accommodate the added weight of the pressure suit helmet, lowering the seat was a must for most pilots. Both the tanker and SR-71 had radio silent procedures to follow, letting each other know when they were ready to commence refueling. Opening of the SR-71's air refueling door meant the SR-71 was ready to receive fuel. Although an experienced SR-71 pilot knew where the contact position was, director lights located on the belly of the KC-135Q gave the pilot clues as to how close he was to the precise refueling position. The boom operator manually controlled the director lights.

Adding a small amount of power, the SR-71 pilot slowly moved forward into the contact position, maneuvering the SR-71 to put the very end of the air refueling boom about three feet outside his front window. Once he stabilized the plane there, the boom operator flew the boom up and back to the awaiting air refueling receptacle. Once the boom nozzle was hydraulically locked into the refueling receptacle, the crews talked to each other over a secure boom interphone system. If there was another tanker in the cell to finalize the

Above: The lead KC-135Q navigator working at his table, figuring out the air refueling rendezvous with the inbound SR-71. By the use of timing, wind drift charts, and experience, the tanker navigator was able to make the rendezvous highly successful in any part of the globe, in all weather conditions. *Lockheed Martin*

Left: The boom operator lies face down on a nylon cot and flies the boom with a control stick, which activates the two black elevons on each side of the boom. He lowers the boom to indicate the SR-71 pilot should move forward. This SR-71 is close to the precontact position, having the boom nozzle about fifty feet in front of the plane. *USAF*

Above: When fuel begins leaking out of the top of the wings, the tanks are full of 80,000 pounds of JP-7. The SR-71's air refueling receptacle door is still open. The pilot will close the door, slide off to one side of the tanker, light the afterburners, and be on his way. *Lockheed Martin*

Right: When the KC-135Q model tankers received the CFM-56 engine in the late 1980s, they were redesignated the KC-135T models. The new engines practically doubled the thrust compared to the original engines. *Lockheed Martin*

refueling, the pilot disconnected from the boom by squeezing the trigger switch on his control stick and maneuvered to the pre-contact position on the second tanker. This was called a "split offload."

The RSO now controlled the fueling of the six main tanks. The goal was to be at the end air refueling (EAR) point with precisely 80,000 pounds of fuel onboard and with a pressure disconnect. Making all three of those requirements happen all at the same time was an art for the RSO. As everyone has experienced when pumping gas into a car's tank at the maximum rate, back pressure will stop the refueling early. More fuel can be obtained if you now pump it slowly. The same principle applied to the SR-71. Transferring fuel into the SR-71 at 6,000 pounds per minute was fast, and once 70 psi backpressure was sensed, the air refueling boom automatically disconnected. That's want you wanted to happen at the EAR point, not prior to reaching it. The RSO had the tankers pump gas using all four of their pumps to begin, but as plane approached the EAR point, the RSO told them to take it down to three, then to two pumps. Within miles of the EAR point,

he had them drop to one pump only. Most RSOs were experts at controlling the tanker's pumps so that the SR-71 arrived precisely at the EAR point with full tanks and a pressure disconnect.

If the mission didn't call for a full load of gas, the pilot could beg for a few extra thousand pounds of gas for insurance. If the tanker crews had extra JP-7, they were always willing to give more. The pilot reminded the tanker aircraft commander which direction he would be turning after refueling, just so there was no confusion. After disconnecting from the boom, the first thing the pilot did was to raise his seat back up to the normal flying position and complete the post-air refueling checklist.

Internal view of one of the six main fuel tanks. The skin of the SR-71 is the fuel tank; there are no internal bladder or metal tanks. The relatively cooler fuel is used as a heat sink for the aircraft's fuselage. *Tony Landis*

Imagine flying on the boom at 350 knots in the dark of night. You're in thick clouds and can barely see the tanker. Air turbulence becomes greater, moving the boom. Distant thunderstorms produce momentary blinding flashes of light. Finally, you have to light one afterburner to stay on the boom. This is why I have gray hair today! *Lockheed Martin*

CHAPTER 14

Acceleration and Climb to Mach 3+ and 71,000 Feet

After safely clearing the tanker cell, it was time to start a simultaneous acceleration and climb to Mach 3+ and 71,000 feet. A transonic region of high drag on the aircraft started just below Mach 0.95 and up to Mach 1.05. The faster the plane accelerated through this region, the better the fuel consumption. The most economical way to get through the high drag speed regime was to do a maneuver called the "Dipsy Doodle." It consisted of a climb and descending acceleration to intercept the SR-71's supersonic climb schedule.

To begin the Dipsy Doodle, the pilot advanced the throttles to minimum afterburner and began a shallow climb at Mach 0.9. Passing through 30,000 feet, the pilot fire walled the throttles to maximum AB. At 33,000 feet he let the airspeed increase to at least 0.95 Mach and slowly lowered the nose to establish and 2,500–3,000 foot-per-minute rate of descent. Passing through the sound barrier at Mach 1.0 was rather benign. The needles of the pitot-static instruments (airspeed, altimeter, and vertical velocity) gave a slight jiggle passing through Mach 1.0, stabilizing quickly. Other than seeing the three needles momentarily fluctuate, there were no other sensations of passing through the sound barrier.

Continuing to accelerate in a shallow dive, the pilot focused on his rate of descent and how rapidly the knot equivalent air speed (KEAS, pronounced "Keys") was increasing. (KEAS is calibrated airspeed corrected for compressibility errors once a plane goes faster than around 300 knots, since air is compressible.) Around 435 KEAS, the pilot began a gradual pull out of the shallow dive and hopefully reached 450 KEAS in climb. Once the airspeed was stabilized at 450 KEAS in the

Opposite: Bright red sunset defines the outline of the SR-71 at full afterburner. *Lockheed Martin*

climb, the pilot engaged the KEAS hold function on the autopilot. The autopilot then held the SR-71 at a constant 450 KEAS climb until reaching Mach 2.6.

If the pilot had not already engaged the automatic navigation system on the autopilot, this was the time to do so. The pilot was able to devote his full attention to more important cockpit indications, switches, and gauges by having the autopilot on. The basic autopilot had two solenoid-held switches, one for the pitch axis and the other for the roll axis. Once engaged, the control stick was disabled, and small, serrated pitch and roll wheels on the autopilot control panel now controlled aircraft movement. The pitch and roll wheels permitted small adjustments by moving the SR-71 approximately one degree in pitch per fifteen degrees of wheel rotation and one degree in roll per eight degrees of rotation.

To aid in making small finger inputs to the wheels, and to keep from tiring out the pressure suit-enclosed right hand, a clear plastic hand rest folded down, in perfect position, to rest the heel of the pilot's hand. He then flew the aircraft by rotating the pitch wheel up and down with his index finger and the roll wheel left and right with his middle finger or ring finger. In this basic autopilot mode, the pilot flew the aircraft through each wheel just as if he had control of the stick. The autopilot had no altitude hold function or auto-throttle feature. Consequently, during cruise, the pilot controlled the aircraft's pitch attitude with the pitch wheel and the airspeed with the throttles.

The forward bypass doors were a rotating band of ports located a short distance aft of the inlet throat. They were located around the circumference of each inlet, at the top and bottom, acting as overboard exhausts for inlet air not required by the engine. They modulated from fully closed to 100 percent open, and were controlled by the air inlet computer (AIC) that measured a comparison of air pressure on the outer surface of the engine inlet to the air pressure inside the inlet. If the air pressure inside the inlet increased, the forward bypass doors modulated open as necessary to relieve excess pressure building up inside the inlet. A comparison would be like trying to drink from a fire hose with the excess water spilling everywhere. At Mach 1.4 the forward bypass doors were barely open. However, by the time the aircraft reached Mach 1.7, they were open as much as 15–20 percent, bypassing excess air overboard.

Passing through Mach 1.4, the forward bypass doors began to open according to inlet pressures. At Mach 1.6, the spikes unlocked and began slowly moving aft into the inlet at the rate of one and five-eighths inch per 0.1 Mach the plane accelerated, up to maximum travel of twenty-six inches aft.

The pilot closely monitored the positioning of the forward bypass doors, the spikes, and the compressor inlet pressure (CIP) gauge. For any given Mach number, there was a specific spike position, altitude, and CIP associated with it. By comparing the actual cockpit readings with the checklist's ideal values, the pilot had

Below: The triple display indicator (TDI) provided meaningful supersonic information for the pilot: 311 knots equivalent airspeed (EAS) at an altitude of 61,100 feet and a Mach number of 1.81. *Rick Poole Photography*

Right: The autopilot controls are on the pilot's right console. You can hand fly the SR-71 at Mach 3+, but to maintain a stable platform the autopilot was used. Engaging the pitch and roll axis of the autopilot disabled the control stick. By use of the two serrated wheels, the pilot flew the plane in pitch and roll with his fingertips. The plastic hand rest, at the bottom, folds forward to provide a base. *Rick Poole Photography*

a good indication of how well the inlets were performing. For example, at Mach 1.7 both the left and right spike indicators should read two inches aft, altitude should be around 43,000 feet and the CIP gauge should be reading 9.0 psi.

The purpose of each inlet spike was to control and position the supersonic airflow inside the throat of the inlet for optimum performance and to prevent supersonic air from entering the engine. One rule true for all aircraft flying supersonic with turbojet engines is that the supersonic airflow and corresponding shock wave must not reach the engine's compressor. If supersonic airflow ever reached an engine's compressor, it would flame out the engine and could possibly damage the compressor blades. The spikes were hydraulically actuated and had to be able to withstand air pressure exceeding fifteen tons under certain airflow conditions. The easiest way to understand how the inlet spike and forward bypass doors operated is to consider the spikes as dumb (they moved fore and aft strictly on a Mach schedule) and the forward bypass doors as smart (they reacted to the existing internal and external air pressures and were controlled by a computer).

There was a major problem with having the forward bypass doors open more than necessary. The excess air exhausted overboard created a tremendous amount of drag on the aircraft as the slower, exiting

Just outboard of the pilot's throttles were the inlet aft bypass controls. The left and right switches were intentionally shaped differently so the pilot could feel, without looking, which aft bypass switch he wanted to activate. Just above them you can see the exhaust gas temperature (EGT) controls that allowed the pilot to increase or decrease each engine's EGT within a given range. *Rick Poole Photography*

On the pilot's right cockpit side were the fuel tank gauge, center of gravity (CG) indicator (with the OFF flag), fuel tank pressure, and liquid nitrogen quantities. An automatic fuel tank sequence maintained the CG, but required the pilot to control it more accurately for maximum performance. By use of the transfer switches, the pilot was able to move fuel to a forward or aft fuel tank to control the CG. *Rick Poole Photography*

air hit the supersonic airstream. In some cases, the drag was significant enough that the mission had to be aborted because of excessive fuel consumption required to overcome the drag. To optimize performance, the computer controlling the forward bypass doors was programmed to run on a relatively tight schedule, keeping the doors closed down as much as possible without creating inlet problems. Crews referred to this as the "inlet schedule." Every SR-71 had a slightly different inlet schedule. Some planes maintained a tighter schedule than others.

At this point in the climb, the aircraft was around 45,000 feet and at Mach 1.7. The forward bypass doors were dumping a significant amount of excess air overboard, creating more drag than necessary. Each inlet system included a second set of doors, called the aft bypass doors, to help close

down the forward bypass doors. Positioning the aft bypass doors was a manual operation by the pilot. Located just outboard of the throttles were individual left and right aft bypass control switches, with fixed positions of either CLOSE (shut), A (15 percent open), B (50 percent open), or OPEN (100 percent open). In future discussions, pay close attention to whether the reference is to the forward or aft bypass doors.

The aft and forward bypass door results worked in opposite direction to each other. If the pilot manually opened the aft bypass doors, the forward bypass doors would tighten down; if he manually closed the aft bypass doors, the forward bypass doors would open up. Think of it in terms of the AIC trying to maintain a constant air pressure inside the inlet. Manually opening up the aft bypass (to let air escape) dictates the forward bypass close down to maintain the same pressure. The advantage was that opening the aft bypass doors to exhaust air did not create any drag on the aircraft. The aft bypass doors kept the excess air inside the inlet, going around the exterior of the engine, cooling it down, and exiting the air in front of the afterburners for additional thrust. As the aircraft reached Mach 1.7, the checklist called for the pilot to manually move the aft bypass doors switches from CLOSE to the A position, allowing the forward bypass doors to close down to around 10 percent.

During the acceleration and climb, the automatic fuel tank sequencing slowly moved the center of gravity (CG) aft. The pilot monitored the CG indicator and pitch trim readings throughout the climb. Since pitch trim worked automatically to keep the aircraft in trim with the autopilot engaged, any unusual readings were a clear indication of an erroneous CG indicator. The RSO had a circular slide rule in the rear cockpit that allowed him to input the fuel readings in each of the six main fuel tanks and arrive at an accurate CG, assuming each of the individual fuel tank readings were accurate.

The inlet guide vanes (IGVs) shifted automatically from an axial position to a cambered position between a compressor inlet temperature (CIT) of 85–115 degrees C., which occurred between Mach 1.7 to 2.3. As each engine's IGVs shifted independently, a noticeable yawing motion of the aircraft took place until both engines shifted to the cambered position. After each IGV shifted, the pilot placed their switches to the LOCKOUT position, preventing the IGVs from shifting back to axial until required later on in the descent. As each IGV shifted, the respective forward bypass door opened to 20 percent or more. The resulting drag required the pilot to place the aft bypass door in the B position (50 percent open). That normally closed the forward bypass doors to around 10 percent open. For the remainder of the acceleration, the forward bypass doors slowly, but surely, closed down.

At this point the climb checklist called for all exterior lights and the pitot heat turned off. Passing through 60,000 feet, the RSO placed the transponder to a nonaltitude reporting position. As far as air traffic controllers were concerned, the SR-71 simply evaporated at 60,000 feet. All the airspace above 60,000 feet, anywhere on the globe, was exclusively ours. By now the plane had accelerated up to around Mach 2.6. The pilot closely monitored the forward bypass doors as they closed down tighter. When the forward bypass doors approached closed, he moved the respective aft bypass door from the B position to the A position.

In a few minutes the SR-71 would level off at Mach 3.0. At Mach 3.0, the maximum airspeed limit was 450 KEAS, and the aircraft was already climbing at 450 KEAS. The pilot needed to gradually reduce the KEAS to be well under the limit. Starting at Mach 2.6, the constant 450 KEAS climb speed the autopilot had been holding began to bleed down automatically by the autopilot. The bleed schedule called for a ten KEAS decrease for each 0.1 Mach number increase above Mach 2.6. At Mach 2.7 the KEAS should be 440, at Mach 2.8 the KEAS should be 430, at Mach 2.9 the KEAS should be 420, and at Mach 3.0, at level off, the KEAS should read 410.

This is the chart the pilot will reference just prior to level off at Mach 3.0. With 50,000 pounds of fuel onboard, he will level off at 71,000 feet, at 407 KEAS, and set each engine's fuel flow to around 20,800 pounds per hour. If the outside air temperature is close to -56.5 degrees C., the cruise chart should be relatively accurate.
Lockheed Martin

Anticipating the level off, the pilot consulted his checklist cruise charts for the Mach 3.0 maximum range profile with around 50,000 pounds of fuel remaining and found the following parameters: altitude 71,000 feet, 407 KEAS, fuel flow for each engine around 21,000 pounds per hour. As 2.97 Mach approached, the pilot disengaged the KEAS hold function of the autopilot and slowly lowered the nose of the plane by rotating the pitch wheel forward with his index finger, leveling off close to 71,000 feet. Simultaneously, he retarded the throttles from the maximum AB climb until each fuel flow gauge read around 21,000 pounds per hour of fuel consumption. This was a good starting point with the throttles to hold Mach 3.0.

The pilot made very small corrections with the pitch wheel and throttles to hold altitude and Mach number. The CIT gauge was the pilot's only indication of whether outside temperatures were hotter or colder than standard day temperatures. If he saw 328 degrees C. on the gauge at Mach 3.0, he knew it was a standard day temperature of -56.5 degrees C. outside; if it was hotter than standard the pilot knew it would most likely cost him more fuel than his computer-generated flight plan stated.

CHAPTER 15

Cruising at Mach 3+

The pilot now had a choice to make, and this is where experience in flying the SR-71 paid off. He had to decide whether to leave the aft bypass doors in the A position, or move them to the CLOSE position. He started by evaluating how the forward bypass doors were modulating at Mach 3.0 cruise. If they were too tight for his comfort level (detailed in chapter 16), he placed the aft bypass in the CLOSE position. If the forward bypass doors were too far open to obtain good fuel economy, he placed the aft bypass doors in the A position. Talking to the crews who recently flew the same plane helped the pilot get an idea if the forward bypass doors were scheduled to run loose or tight. Unfortunately, each SR-71 often had a mind of its own. Anytime the SR-71 was flown at or above Mach 3.05, the aft bypass was always placed in the CLOSE position.

Cruising at Mach 3+ the airspace was all ours. Because of the tremendous rate of fuel consumption in afterburner cruise (around 42,000 pounds per hour initially), pilots flew continuously at the optimum cruise altitude for the aircraft as gross weight decreased. That worked out to about a 100 to 150 foot per minute rate of climb, which kept the aircraft at its optimum altitude for fuel efficiency. Temperature deviations ("temp devs") from standard day temperatures would either aid or hinder aircraft performance. At hotter than standard day temperatures, the inlets were not as efficient because the forward bypass doors opened more and required additional thrust during cruise. As an example, a twenty-degree temperature increase will increase total fuel flow an additional 9,000 pounds per hour: a considerable amount!

The cruise leg was where experience in the SR-71 really paid off. After three to four hundred hours in the Blackbird, it became instinctive where to position the forward and aft bypass doors, achieving optimum fuel efficiency. The new pilot learned by listening to the old heads and trial and error. The name of the game was being able to achieve the best cruise performance in any part of the world for that particular plane. Former RSO Col. Frank Stampf (Ret) explains how exceptional aircraft performance can get you into trouble you never expected.

Extreme temperatures at altitude could affect the SR-71 in more ways than one. In addition to much warmer than standard temperatures causing Habu crews lots of problems by robbing them of precious JP-7 fuel, significantly colder than standard temperatures gave us a whole set of challenges.

Opposite: By raising the outboard section of the wing, maintenance personnel had full access to the engine. The small, whitish cylinder, on top of the engine contains 1¼ pint of triethylborane (TEB) for ignition of the main and afterburner fuel. *Tony Landis*

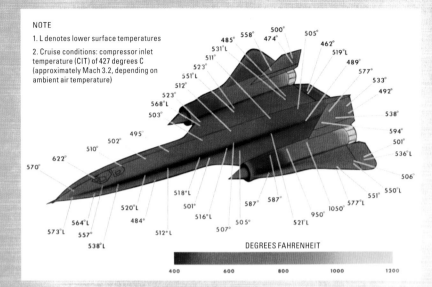

NOTE

1. L denotes lower surface temperatures

2. Cruise conditions: compressor inlet temperature (CIT) of 427 degrees C (approximately Mach 3.2, depending on ambient air temperature)

500°
558°
485°
474°
505°
531°L
462°
511°
519°L
523°
489°
551°L
577°
512°
533°
523°
492°
568°L
503°
538°
495°
594°
502°
501°
510°
536°L
622°
570°
506°
550°L
518°L
551°
520°L
501°
587°
587°
577°L
484°
516°L
1050°
950°
521°L
573°L
564°L
557°
512°L
507°
505°
538°L

DEGREES FAHRENHEIT

400   600   800   1000   1200

These are the approximate exterior temperatures while flying at Mach 3.2. The amazing thing is, these temperatures are created where the outside air temperature is at -70 degrees F. Because of the heating and cooling cycles the SR-71 went through, the airframe actually became stronger over time by the process of annealing, the same process blacksmiths use on horseshoes to make them stronger. *Lockheed Martin*

Gil, my pilot, and I were flying a "routine" sortie out of RAF Mildenhall into the Barents Sea. I don't remember being briefed about any specific weather problems for the track. Our mission was right out of the textbook as we broke off from our tankers and started the acceleration that would take us "round the bend" of the Kola Peninsula. The fact that things were going so smoothly up to that point should have made us suspect that something out of the ordinary was probably about to happen!

Not long into the acceleration, both Gil and I remarked that the airplane was climbing like the proverbial "bat out of Hell." The outside temperatures were quite a bit colder than anticipated, and that always made a difference in performance. I don't think either of us were expecting the ride we were about to get. Gil started to ease the throttles back to level off at Mach 3.0. However, although we were ready to level off, the airplane wasn't!

The outside air temperatures were running twenty-nine degrees below standard, almost twice as cold as either of us had ever experienced in our relatively short SR-71 flying careers. The result was that the airplane just wanted to keep climbing and accelerating past Mach 3.0 and 80,000 feet.

To most aircrews flying more "normal" aircraft, that probably sounds pretty neat; unfortunately it put us in a bit of a predicament. If Gil let the airplane continue its climb on its own, we'd soon bump up against the minimum allowable airspeed of 310 KEAS—avoiding it was critical to providing aerodynamic controllability— and we were losing speed rapidly as our altitude was increasing. If we leveled off to maintain minimum allowable KEAS, the resultant higher than planned Mach number would more than likely force us outside of our preplanned turn radius. Our closest point of approach to Soviet defenses was close enough without that kind of help!

Since aborting the mission because the airplane was performing better than we'd expected seemed a bit embarrassing, we

accepted the high Mach, and Gil manually increased our bank angles to keep us on our planned track through the sensitive area. This mission was successful, and we learned yet another lesson about just how many different ways the SR-71 could surprise us.

If the forward bypass doors were opened too far, the increased drag decreased aircraft performance, potentially to the point of having to abort the mission for insufficient fuel. However, keeping the forward bypass doors opened more than optimum was the conservative way to keep from encountering the dreaded "unstarts" from occurring (see chapter 16). An unstart and its associated recovery process would put the aircraft further behind on fuel consumption. It was hard to win!

At Mach 3.0, it often worked out that putting the aft bypass doors in the CLOSE position opened the forward bypass doors too much, and putting the aft bypass doors in the A position closed them down too tightly. With this scenario, the best solution was to keep the aft bypass doors in the CLOSE position and merely fly the SR-71 slightly faster, around Mach 3.05. By flying at a slightly higher Mach number, the forward bypass doors typically closed down to an optimum setting. If the forward bypass doors modulated between 5 and 10 percent open, drag was minimized and aircraft performance was optimized.

Although crews flew maximum range cruise profiles nearly 98 percent of the time, they had to be prepared to fly the following profiles as well:

Cruising at 71,000 feet. Looking out one window there is not much curvature of the Earth, but when you look from side to side there is a definite curvature. The sky is dark blue on the horizon and becomes completely black at about forty-five degrees above the horizon line. Stars are visible in the black sky during the daytime. *USAF*

Top left: The "working end" of the J-58 engine. The four concentric rings are the fuel nozzles. Where the pilot placed the throttles in the AB range dictated how many fuel nozzles were spraying raw fuel into the AB section. In full AB, all four nozzles provide maximum fuel. The whitish lining is made of ceramic to help keep the AB section cool. *Tony Landis*

Top right: At the far left side of the picture, you can see the installed engine's afterburner section. Everything behind the afterburner is part of the aircraft. The engine afterburner nozzle is enclosed by a fixed-contour, convergent-divergent ejector nozzle, to which free floating trailing edge flaps are attached (right side). The trailing edge flaps are free to open and close with varying internal nozzle pressure. Maintenance nicknamed them "Turkey Feathers." *Tony Landis*

Right: While visiting personnel at RAF Mildenhall, one of the SR-71s had major engine problems and diverted into Bodo, Norway. I flew a KC-135Q up to Bodo to help with the recovery, and surveyed the damage in one of their hangars. The plane needed a complete engine change, requiring a C-141 to bring in the new engine. *Rich Graham*

A. **Minimum AB cruise:** This profile yielded the lowest cruise altitude for the Mach specified and usually resulted in less than maximum range.

B. **Maximum range (Optimum cruise):** This profile yielded maximum range for the Mach specified. Power settings were in the lower portion of the AB throttle range.

C. **Intermediate altitude cruise:** This profile yielded altitudes above maximum range profile but below the maximum altitude cruise profile. Specific range was less than optimum but reasonable.

D. **Maximum altitude cruise:** This profile resulted in altitude approximately 1,000 feet below the maximum AB ceiling for the Mach specified.

E. **Maximum AB ceiling:** This profile required continuous operation at maximum AB and the Mach specified. This is normally 4,000–5,000 feet above the altitude schedule for maximum range. A descent of 2,000 feet had to be completed prior to any turn entry.

The design Mach of the SR–71 was Mach 3.2. The maximum altitude limit was 85,000 feet unless specifically authorized to fly higher. The J-58 engine

ONE OF MY MOST MEMORABLE SIGHTS was during a flight from Beale to RAF Mildenhall. We finished our refueling over Goose Bay and had leveled off at Mach 3.0 heading out over the North Atlantic. It was pitch black outside when the sun began to rise rapidly off the nose of the aircraft. Watching a spectacular sunrise in front of the aircraft, I peeked in the periscope just in time to see a brilliant full moon behind me, centered precisely between the two blue afterburner exhausts with their associated concentric shock waves inside. It was a true "photo-op" moment!

Some of the most notable sights were at night crossing the North Atlantic enroute to Mildenhall. The Northern Lights, viewed from 75,000 feet on a clear night, were a spectacle of vivid green and blue colors, moving and dancing outside both windows. They surrounded the aircraft in what appeared to be colorful sheets and staying with us forever, before rapidly disappearing as the sun began to rise in the east. At night all the stars appeared to be three-dimensional, as if you could actually reach out and touch them. Impossible, but a wonderful illusion!

Sunset with the SR-71 in full AB and the concentric shock waves. At Mach 3+ cruise at night, looking through the periscope, the AB appears like Bunsen burner flames, very pointed and dark blue in color. *Lockheed Martin*

is the only turbojet engine ever certified for running continuously in AB. Most jet aircraft have restrictions on how long they can remain in AB. Performance charts provided three power settings for cruising in AB. Flying the maximum altitude profile required the throttle to be near the full AB stops. Flying at maximum range most of the time placed the throttles at the lowest end of the AB range, sometimes coming out of AB inadvertently. Flying at the intermediate altitude cruise had the throttles about half way in the AB range.

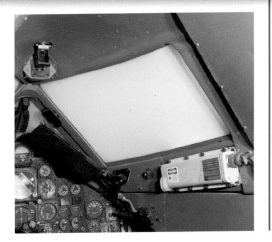

The SR-71 was not created for night reconnaissance. However, once a high-quality radar imaging system was developed, night flying became routine. Several crews came close to losing the plane at night with unstarts and steep banks. To aid pilots in their orientation to the horizon at night a peripheral vision device (at right) was developed. This projected a gyro-stabilized, red laser line across the cockpit front panel, representing the horizon. *Rick Poole Photography*

CHAPTER 16

Inlet Unstarts and Restarts

In order to understand unstarts, you first have to understand the air inlet system itself. When most aviators think of supersonic jets, they focus on the tremendous thrust the aircraft's engine(s) puts out to reach such high speeds. In the case of the SR-71, it was not the 68,000 pounds of thrust the two Pratt & Whitney J-58s produced that allowed the aircraft to attain speeds in excess of Mach 3.0—it was the inlets. Kelly Johnson and other Skunk Works engineers, like Ben Rich, knew if they could somehow capture and harness the high-speed airflow going over and around the SR-71, it could be used to provide additional thrust. Kelly Johnson was often quoted as saying, "At Mach 3.2 cruise [the design Mach number], the J-58 engines produced approximately 20 percent of the thrust, while the inlet produced around 80 percent of the thrust." Arnie Gunderson, who worked for Pratt & Whitney and is known in Blackbird circles as "Mr. J-58," was the technical representative manager of all the SR-71's engines. He sheds light on how the SR-71 gained additional thrust by utilizing the Mach 3.2 air entering the inlet.

What is ram air? It is an aspect of airflow over and through a moving object. We know that air creates drag or friction as it flows over the airplane. But when the flow goes into an opening in the airplane, such as an engine inlet, and is slowed down, its static pressure increases. Think about sticking your hand out the window of a car while it is moving at highway speeds. You hold your hand flat, like a wing, and can move it up and down with slight movements of your wrist. That shows lift. But the force that is pulling your hand backward is called drag. Now close your hand into a fist. The only thing you feel is drag. Open your hand into a cup shape to capture the air. The force in your palm can be called ram air. The air is slowed down (relative to your hand), and the pressure has increased.

The same thing happens in the nacelle of the SR-71. Air enters the inlet and is slowed down to about 0.6 Mach (relative to its Mach 3.2 entry speed). It gains a lot of pressure and temperature in that process. That's energy that can be recovered and

Opposite: Inflight shot of the "Big Tail" modification. The addition of the long tail detracted from the graceful lines of the SR-71. *Lockheed Martin*

used, if you're smart. The SR-71 nacelle flows that air around and through the engine, and then it is finally expanded through the exhaust ejector at the very end of the nacelle. That final expansion recovers about 88 percent of the original energy imparted to the air as it first entered the nacelle a few milliseconds earlier. So the bottom line for the SR-71 is not

ram, but ram recovery. In a few words, ram recovery is capturing the supersonic shock wave and using it to create additional thrust.

The term "unstart" implies that something must be started to begin with. The conical supersonic shock wave formed at the very tip of each spike and

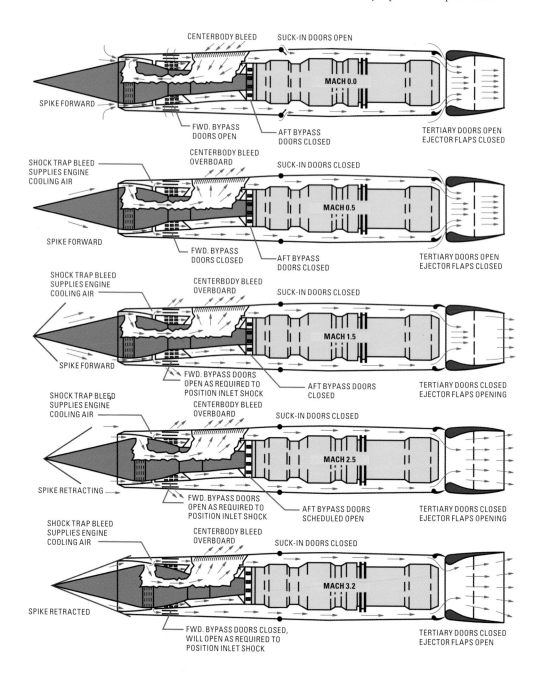

Five different airflow patterns inside the inlet at increasing Mach numbers. Around 80 percent of the total thrust comes from the inlet geometry and capturing the Mach 3 air to augment the afterburner thrust. The actual J-58 engine produces the other 20 percent of the thrust. *Lockheed Martin*

narrowed down the faster the aircraft went, eventually entering the inlet. When the shock wave first entered the inlet, that inlet was considered to be "started." The inlets usually started between Mach 1.6 and Mach 1.8. An unstart occurred when the shock wave was rapidly ejected back outside the inlet. Anytime the inlet air pressure became unacceptable, a phenomenon known as an aerodynamic disturbance, or unstart, took place. Crew members shortened aerodynamic disturbance to "AD" and used the terms "unstart" and "AD" interchangeably during discussions. The air inlet computer maintained very tight air pressure parameters inside the inlet.

The expulsion of the supersonic shock wave inside the inlet to outside occurred in a fraction of a second. The inlet was no longer started. Unstarts are a phenomena of airflow in a supersonic inlet when the inlet pressure behind the shock wave becomes too great or the spike position is too far aft. Improper spike and/or forward bypass door position could result from inlet control errors, computer failures, loss of hydraulic pressure, electrical glitches, or mechanical failures. Any of these malfunctions could cause an unstart.

From a crew member's perspective, unstarts were characterized by a loud "bang," airframe roughness, aircraft yawing and rolling, and a distinct increase in drag and loss of thrust. The J-58 engines continued to run; however the AB might have blown out. The inlet instruments showed a rapid decrease in the CIP to 4 psi, the EGT rose, and fuel flow decreased. All of these events happened instantly as the shock wave was expelled from the inlet. The aircraft yawed and rolled rapidly toward the unstarted inlet. A pitch-up tendency occurred due to yaw and roll rates developing during the unstart. During turns, roll control was critical if the unstart occurred on the down-wing inlet. An unstart was usually unpredictable and violent, mainly because the inlets and engines were located so far away from the centerline of the aircraft.

Above: Congratulations from the Det 1 Commander, Lee Shelton (blue uniform). On 26 October 1987 the crew of Mac McKendree (left) and Randy Shelhorse completed a grueling 10.8 hour mission into the Gulf area from Okinawa, requiring five air refuelings. *Mac McKendree*

Left: Maintenance technician working in one of the chine bays housing electronic equipment. Originally, all the bays were insulated with gold foil, the best thing available. Over time, the gold was replaced with cheaper insulating materials. *Tony Landis*

Unstarts generally occurred when pilots least expected them—all relaxed and taking in the magnificent view from 75,000 feet. The instant an unstart occurred, the first step the pilot was trained in the simulator to do was to keep the angle of attack within limits. In other words, don't let the aircraft pitch up, and hold the wings as level as possible. An experienced pilot could sometimes feel that an unstart was about to occur if the forward bypass door was closing down too tightly and a very subtle inlet duct rumble manifested itself throughout the airframe.

Some of the early A-12s were lost in the 1960s, as pilots experienced unstarts for the very first time. Working closely with test pilots, the Skunk Works engineers began to understand unstarts better and came up with a solution to minimize the extreme yawing they created. The inlets incorporated an automatic "restart" feature to aid the pilots recover from a violet unstart. They had to deal with both an immediate large loss of thrust during an unstart and a tremendous amount of drag created by the automatic restart cycle. To keep the drag as symmetrical as possible

during restart, the engineers came up with a cross-tie system. This was enabled when the aircraft accelerated through Mach 2.3. If an unstart occurred above Mach 2.3, the cross-tie feature sympathetically restarted the good engine as well. This helped to minimize the extreme rolling and yawing of the aircraft and prevented the good inlet from unstarting as well. The downside to the cross-tie feature was that total drag instantly doubled, as both inlets went through the automatic restart cycle. This was regarded as the lesser of two evils.

An inlet control system called the shock expulsion sensor (SES) used CIP as its reference since a rapid CIP decrease was characteristic of an impending unstart. Whenever the CIP rapidly dropped, the SES actuated and initiated the automatic restart cycle on the offending inlet, or both inlets if the plane was above Mach 2.3.

A state-of-the-art DEF H defensive radar jammer, one of the most powerful of its time, sits on its loading dolly ready to be uploaded into the insulated bay underneath the SR-71. The silver/green unit on the left is its evaporative cooling system. *Mike Hull*

To recapture the supersonic shock wave and airflow after an unstart above Mach 2.3, the automatic restart cycle went through the following sequence in about twenty seconds on both inlets: the forward bypass doors went to full open; the spikes rapidly moved forward as much as fifteen inches to recapture the shock wave; the spikes then retracted to their scheduled position; then the forward bypass door returned to automatic operation and began to close down to satisfy the amount of air to bypass. If the unstart didn't correct itself after one automatic restart cycle, the cycle would repeat itself until it affected a restart or the pilot took control of the inlet manually.

During an unstart, the pilot's primary concern was aircraft control. Once that was taken care of, he had to figure out which inlet had unstarted and why in order to apply corrective actions on the offending inlet. Below Mach 2.3, it was obvious. Above Mach 2.3, with cross-tie active, the pilot could tell by watching the spike and door indicator needles on the instrument panel. Whichever spike and door needle was moving to its restart position first was the offending inlet; the lagging needle was the good inlet, sympathetically restarting. The initial yawing motion was another clue.

Depending on the position of their heads at the time of an unstart, a few crew members actually cracked their pressure suit helmets by hitting them on the side of canopy from the force of the unstart. The vast majority of unstarts cleared themselves after one automatic restart cycle and were considered transient events. The crew pressed on with the mission and paid more attention to the inlet parameters. Over the history of the Blackbird program, many improvements reduced the severity and frequency of unstarts and provided a better cockpit presentation of what was going on inside the inlets. In the mid-1980s a major modernization effort was accomplished on the entire SR-71 fleet, called the digital automatic flight and inlet control system (DAFICS). Until the DAFICS upgrade, the entire SR-71 fleet was running on analog

FLIGHT TO FREEDOM
by
Robert Lorenz

Victor Belenko who in 1976 escaped from the
Soviet Union flying a MiG-25 (Foxbat)
now flies demonstration flights for the
BEDE Jet Corporation.

computers and flight controls. After the upgrade, everything changed to digital control, improving the reliability and maintainability of the aircraft. As a result of DAFICS, unstarts were minimized, but never completely eliminated. (Whenever I do something stupid my wife just shakes her head and murmurs, "Poor guy . . . too many unstarts.")

No SR-71 has ever been lost or damaged due to hostile actions. The aircraft was extremely difficult for enemy radars to find. Featuring the original stealth technology, the SR-71's wings, tail, and fuselage were coated with paint containing iron ferrites to absorb radar energy rather than returning it to the sender. In effect, the SR-71 became the first stealth plane. At cruise speeds and altitude, the radar cross section of the SR-71 represented a target about the size of a J-3 Piper Cub. Even if the SR-71 could be found on radar, its detection was so late that there was simply not enough time for a SAM to compute the lead for a successful kill. The plane also carried sophisticated electronic countermeasures (ECM) equipment to electronically jam and deny the enemy our range and bearing. Consequently, the SR-71's ECM was continually updated to counter the latest enemy electronic order of

Victor Belenko and the Mig-25 he was flying while trying to intercept the SR-71. Victor is now a U.S. citizen who I've met and talked with at the EAA air show at Oshkosh. At the bottom of the picture is a BD-10 prototype aircraft that Victor has also flown.
Rich Graham

At the top center of the canopy is the rear-view periscope. When the white nylon handle is pushed to the up position (as seen), the viewing tube is extended into the airstream. The cone of viewing is around thirty degrees total angle aft. The two small mirrors on the front canopy sill are used as an aid for the pilot to find items on his pressure suit. *Rick Poole Photography*

side-angle cameras. They taunted and toyed with the MiG-25s sent up to intercept them, scooting up to altitudes the Soviet planes could not reach, and circling leisurely above them or dashing off at speeds the Russians could not match.

[The Soviets] had a master plan to intercept an SR-71 by positioning a MiG-25 in front of it and one below it, and when the SR-71 passed they would fire missiles. But it never occurred. Soviet computers were very primitive, and there is no way that mission can be accomplished.

First of all, the SR-71 flies too high and too fast. The MiG-25 cannot reach it or catch it. Secondly . . . the missiles are useless above 27,000 meters [88,000 feet], and as you know, the SR-71 cruises much higher. But even if we could reach it, our missiles lack the velocity to overtake the SR-71 if they are fired in a tail chase. And if they are fired head-on, the guidance systems cannot adjust quickly enough to the high closing speed.

Another reason Belenko's missiles wouldn't work against the SR-71 is that most air-to-air missiles are optimized to maneuver in the thicker air below around 30,000 feet in order to shoot down an enemy plane. Firing at the SR-71, cruising at 75,000 feet, the air is so thin that any maneuvering capability of the missile is practically nonexistent.

As the SR-71 entered each turn, the pilot increased each throttle an additional 3,000 pounds per hour on the engine's fuel flow gauge to hold the Mach. Constant altitude turns, up to thirty-five degrees of bank, could normally be made at maximum cruise range altitudes by merely adding power. During steep banked turns (more than thirty-five degrees), it was not always possible to maintain cruise altitude. Whenever the aircraft became thrust limited during a steep banked turn, it was better to lose altitude to maintain the Mach, than to lose Mach and maintain altitude. Pilots anticipated and evaluated the possible

battle (EOB). The EOB for flying in Europe was considerably different than flying in the Pacific region. The RSO could control the jammers manually or let them run in automatic mode to electronically jam a specific threat frequency.

On 6 September 1976, Soviet pilot Lt. Viktor Belenko defected to Japan in a MiG-25. Subsequently, he wrote a book titled *MiG Pilot* in which he described the intercept problem.

American reconnaissance planes, SR-71s, were prowling off the coast, staying outside Soviet airspace by photographing terrain hundreds of miles inland with

IT STILL AMAZES ME EVEN TODAY that astronomers have charted our galaxy so accurately that it allowed the ANS to calculate the SR-71's position so precisely. Things may have changed here on Earth from century to century, but the same stars guided both Christopher Columbus and Habus.

The ANS controls were only in the RSO's cockpit. The pilot's instrument panel had all the navigational displays, but no actual control of the ANS itself. Obviously, many of our worldwide reconnaissance missions were not in range of any friendly navigational aids. If for some reason the RSO were incapacitated inflight, the pilot had no way of telling the aircraft where to navigate to. The best the pilot could do was to let the ANS continue navigating or disengage the automatic

navigation and steer the plane manually in a direction towards friendly airspace, hoping to find an airfield to land on.

On the other side of the coin, the RSO had no flight controls or throttles in his cockpit. If the pilot became incapacitated inflight, the best option for the RSO would be to direct the ANS to turn the plane towards friendly airspace, hope for the best, and eject once the plane ran out of fuel.

It was an automatic abort if the ANS wasn't working properly. My RSO, Don Emmons, had total responsibility for making abort decisions concerning our navigational accuracy. Fortunately, this never happened during our seven years of flying the SR-71 together.

Det 1 crews, wearing their "orange bags," and staff personnel in front of an SR-71 hangar around 1977. Left to right: Rich Graham, Bill Groninger, Don Emmons, Dave Nicola (maintenance officer), Jay Murphy (director of operations), George Chumbley (intelligence officer), Chuck Sober, Barry MacKean, and Lee Shelton. *Rich Graham*

The OBC. The bulbous nose
assembly of the camera
rotates continuously while
in operation. In the very back
of the camera are the film
supply and take-up spools.
The camera was capable of
continuous film coverage
thirty-six nautical miles left
and right of the aircraft track.
USAF

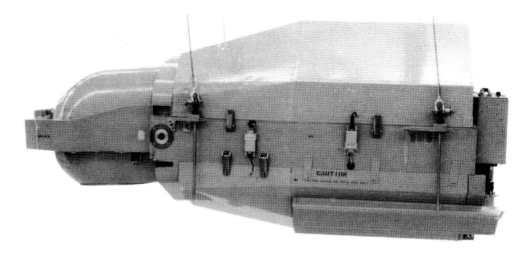

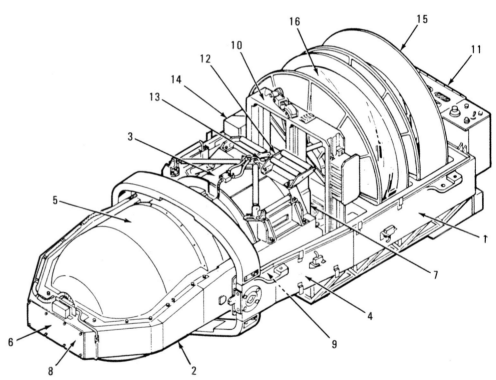

OPTICAL BAR CAMERA

NO.	REFERENCE DESIGNATION	NOMENCLATURE	NO.	REFERENCE DESIGNATION	NOMENCLATURE
1	A1	Main frame	9	A1A1A10	Input Drive (Far Side)
2	A1A1	Gimbal Assembly	10	A1A2	Shuttle Assembly
3	A1A1A1	Lens and Roll Gimbal	11	A1A4	Electronic System Assembly
4	A1A1A2	Framing Roller	12	A1A12	Metering Roll Assembly
5	A1A1A3	Lens Cover	13	A1A11	Stereo Drive Assembly
6	A1A1A4	Slip Ring	14	A1A16	Power Supply and Filters
7	A1A1A5	Auxiliary Data	15	A1A17	Takeup Spool Assembly
8	A1A1A7	Encoder	16	A1A20	Supply Spool Assembly

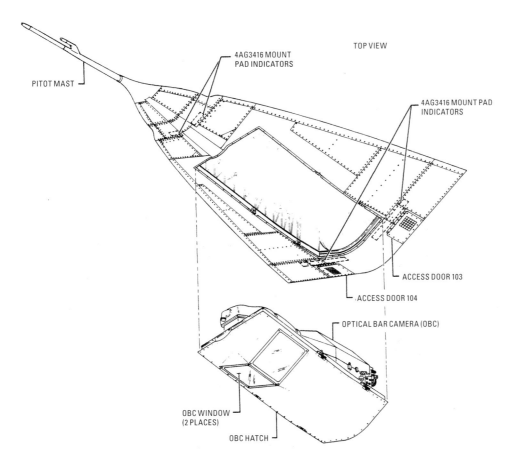

TOP VIEW

PITOT MAST

4AG3416 MOUNT
PAD INDICATORS

4AG3416 MOUNT PAD
INDICATORS

ACCESS DOOR 103

ACCESS DOOR 104

OPTICAL BAR CAMERA (OBC)

OBC WINDOW
(2 PLACES)

OBC HATCH

Left: The OBC is located in the nose section. The two OBC windows are kept at 104 degrees F by a ground heating unit until the aircraft has its engines started and is running its own internal heating system. Precise temperature of the windows was vital to obtain good imagery. *USAF*

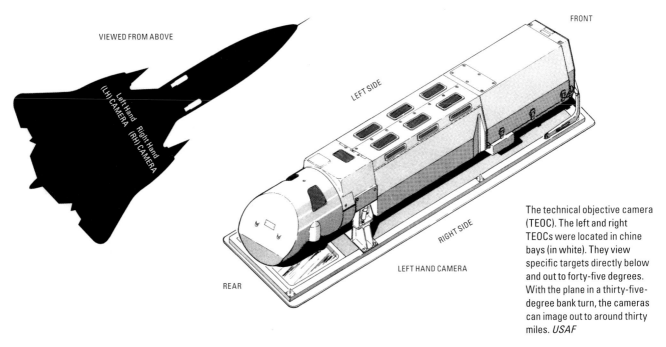

VIEWED FROM ABOVE

Left Hand
(LH) CAMERA

Right Hand
(RH) CAMERA

FRONT

LEFT SIDE

RIGHT SIDE

REAR

LEFT HAND CAMERA

The technical objective camera (TEOC). The left and right TEOCs were located in chine bays (in white). They view specific targets directly below and out to forty-five degrees. With the plane in a thirty-five-degree bank turn, the cameras can image out to around thirty miles. *USAF*

need for a descent prior to every forty-five-degree banked turn, particularly if temp devs were hotter than standard.

Each inlet was biased by the AIC to a slightly more conservative spike and door position during turns, reducing the possibility of unstarts. Each inlet was specifically biased for increased load factor (g-loading), increased AOA, and sideslip (yaw). As the aircraft entered each turn, the forward bypass doors were biased slight more open and the spikes moved forward a small amount. If the pilot saw the forward bypass doors opened further than he liked during the turn, he placed the aft bypass door to the A position, closing down the forward bypass doors and increasing fuel efficiency throughout the turn. There was one problem with using this technique: The pilot had to pay close attention to when the autopilot was about to roll out of the turn. If he didn't move the aft bypass back to the CLOSE position during the rollout, he was setting himself up for a possible unstart.

After the initial level off, it was a good time to look in the periscope to see if the

rudders were centered. Yes, as hard as it is to believe, the pilot actually had a periscope mounted in the top of his canopy. It was moved into the airstream by pushing upward on a white handle mounted on the aft side of the viewing tube. The rearward cone of view was approximately ten degrees across; however, head movement by the pilot extended the viewing cone to around thirty degrees total. The lens produced a two-to-one reduction ratio. The pilot was able to trim up each rudder so they were streamlined with the fixed base of the rudder, mounted on each engine nacelle. That minimized any trim drag on the plane for increased efficiency. It was necessary to check the rudders periodically throughout the mission to make sure they were still aligned. Heating of the hydraulic fluid and rudder actuators would often cause the rudders to become misaligned. The periscope was also used to verify external engine damage or fire, check for the presence of a contrail showing the SR-71's flight path, and confirm if fuel dumping was on or off.

The SR-71's high speed and politically sensitive missions demanded a navigational

Inside the TEOC. The camera had a forty-eight-inch focal length with an optical resolution down to six inches or better, from 80,000 feet, depending on weather conditions. A story was told about photographing a golf ball on a putting green from 70,000 feet . . . if it's true, it had to be perfect atmospheric conditions. *USAF*

SYSTEM SCHEMATIC

OBLIQUE HEAD   LENS CONE MOUNT   SUPPORT HOUSING   MAGAZINE

IR AND HAZE FILTER

FMC ACTUATOR   OBLIQUE DRIVE

UNCLASSIFIED

FLIGHT

DATA CHAMBERS

SHUTTER   PLATEN   CARRIAGE   FRONT STABILIZER*

ELECTRONICS

ELECTRONICS   ELECTRONICS

MIRROR   LENS   REAR STABILIZER*   LENS SUPPORT TUBE   VACUUM VALVE   SHUTTLE   FORWARD END

WINDOW   CAM SWITCH

*HR-308C MODEL SHOWN
HR-308B HAS PASSIVE ISOLATORS

system that was highly accurate and reliable. Further, it needed to be independent, not relying on ground sources for information, which were subject to electronic jamming. The SR-71's astro-inertial navigation system (ANS) filled all of those requirements. Simplistically, the ANS was a star-tracking navigation system. At least two different stars had to be tracked for optimum navigational performance. With a highly accurate chronometer (to the hundredth of a second) supplying the time and the Julian date, along with a catalog of sixty-one of the brightest stars in the northern hemisphere stored inside the ANS computer, it was possible to know precisely where the SR-71 was over the ground.

Selection of which star to locate and track was made by the ANS computer as a function of latitude, longitude, day of the year, time of day, aircraft pitch and roll, and location of the sun. The computer selected a star by going through its star catalog, which was arranged in decreasing star brightness. A telescope-like star tracker looked for the stars in an expanding rectangular, spiral search pattern. The ANS window was located on top of the fuselage, just forward of the air refueling door, and consisted of a round piece of distortion-free, high-quality quartz glass (about nine inches diameter) that allowed the star tracker to see through. Star tracking was automatic.

On the right side of the RSO's cockpit was the ANS control panel. A star

In an attempt to get additional sensor and jamming capability onboard the SR-71, a test version of the "Big Tail" modification was added to aircraft 959. Because the tail would scrape the ground on takeoff and landing, it retracted upward. It proved feasible, but the test program was short-lived. *Lockheed Martin*

The RC-135 reconnaissance aircraft. It seats more than thirty people, including the cockpit crew, electronic warfare officer, intelligence operators and in-flight maintenance technicians. The 55th Wing at Offutt AFB, Nebraska, is home for the RC-135s, but various forward deployment locations are used for worldwide operations. The many unusual antennas along the exterior are for gathering intelligence. *Key Publishing*

"ON" light illuminated indicating that a minimum of two different stars had been tracked within the last five minutes. By comparing the position of the stars to their known location, and with the exact time of day, the ANS could then compute the aircraft's precise position. The ANS provided the SR–71 with an accuracy of 1,885 feet (0.3 NM) for up to ten hours of flying time. Once on course, the ANS provided 300 foot accuracy either side of the programmed track.

The heart of the ANS was a large, self-contained unit—about half the size of a large refrigerator—called the guidance group. A computer within calculated auto-navigation, guidance, and avionics control and maintained a continuously updated account of navigational status and coordinate values. The computer also stored instrument and mathematical coefficients, predetermined data references that defined the stars, and the mission flight plan. Software corrections to the star data were provided for the supersonic shock wave over

the star tracker window that refracted the star light and for pressure and temperature gradients acting on the window, causing optical lens effects.

The aircraft's flight plan and sensor operations for the entire mission were contained on a wide tape made by the mission planners, which was punched with a series of holes and loaded inside the guidance group computer memory. The guidance group orchestrated everything concerning track and sensor operations. It was so smart that it acquired the nickname of "R2-D2," after the navigation droid in *Star Wars*. As the tape ran inside the guidance group, the pattern of holes "told" the aircraft where to navigate, what bank angle for turns, when various sensors were to turn on or off, and where to have the sensor "look" for intelligence gathering. Prior to every flight, ANS maintenance personnel loaded the tape and ran the guidance group in their shop to insure the programming was correct. It was delivered to the aircraft several hours before takeoff, hoisted up by a crane assembly and slowly lowered into its air conditioned bay located directly in front of the air refueling door. An exterior aircraft panel containing the star tracker window bolted over the guidance group bay.

The primary imaging sensor for the SR-71 was located in the nose of the aircraft, either a photographic camera or a radar-imaging sensor. The photographic camera, the optical bar camera (OBC), utilized a continuous moving roll of film taking pictures 36 NM on each side of the aircraft. Sufficient film was loaded to cover approximately 2,952 NM, or half that distance in stereo mode. Around 1986, a new state-of-the-art, high-resolution radar imaging system, called the advanced synthetic aperture radar system (ASARS), was developed. With the advent of ASARS, the SR-71 was now a highly capable, day/night, all-weather, reconnaissance platform. The ASARS had the capability for search, acquisition (navigation updates), and two high-resolution spotlight modes. In the search and spotlight modes, the imaged area was perpendicular to the SR-71's ground track, or the system could be aimed forward or aft up to thirty degrees. In the spotlight mode of operation, the image resolution was down to one foot.

The SR-71 was always equipped with high-resolution, technical objective cameras (TEOCs) for imaging specific targets areas. These were installed in the left and right mission bays underneath the fuselage. The ANS automatically controlled the TEOC operation for specific targets on the ground. Each TEOC had enough film for around 1,428 NM of coverage. Resolution was down to a matter of inches.

Another form of intelligence gathering is called ELINT, "electronic intelligence"— the recording of electronic signals from the low end of the frequency spectrum (e.g., radar acquisition, tracking and guidance signals) to the very high frequencies of the Soviet SA-10 missile. The SR-71 was excellent at stimulating the enemy's electronic environment. Any time an SR-71 flew near or over a foreign country, all kinds of radars and other electronic wizardry were turned on to see if they could find out what was flying so fast within their airspace. Electromagnetic reconnaissance systems received and recorded all their signals at a distance over 650 NM, both left and right of the SR-71 position. The system often recorded over five hundred emitters on a single sortie out of Mildenhall because of the large number of Soviet and Warsaw Pact electronic emitters. It was a *Star Wars* version of eavesdropping.

Although the SR-71 was good at gathering ELINT, other reconnaissance assets were even better. The 55th Wing at Offutt AFB, Nebraska, flew the RC-135 aircraft, with its own unique capabilities for gleaning valuable electronic intelligence. SR-71 crew sometimes flew sorties in concert with the RC-135s, known as "coordinated" missions. On coordinated sorties, the SR-71 not only gathered intelligence, but also stimulated the electronic environment of the foreign country of interest, primarily the former Soviet Union.

Coordinated sorties were timed so that the RC-135 would be flying well outside the borders of a foreign country at about 30,000 feet with all its sophisticated eavesdropping electronic equipment up and running, while the SR-71 flew at Mach 3+ much closer to the border. Many early warning radar and missile radar operators attempted to locate, track, and lock onto the SR-71, while far off in the distance the RC-135 was recording all their electronic signals. One coordinated mission flown in 1978 out of Mildenhall with an RC-135 was the first time significant ELINT was ever recorded of the Soviet's SA-5 missile. At Det 4, the SR-71 also flew coordinated sorties with the British RAF Nimrod aircraft and the German Atlantiques aircraft, gathering ELINT. Once the electronic signature of enemy radar was uncovered, the U.S. could then develop jamming systems to defeat the radar's capability.

CHAPTER 17

Entering the Sensitive Area

As the SR-71 approached the sensitive area, each crew member became more focused, gaining an increased sense of awareness. The sensitive area was airspace where anything and everything was possible. It might be Soviet MiGs trying to intercept the plane or SAMs launched in an attempt to shoot down the SR-71. It could also be where nothing at all happened—you just never knew. Time in the sensitive area varied with every mission but was typically between five to fifteen minutes.

Prior to entering the sensitive area, the crew completed another checklist. The first item was disarming the parachute emergency locator beacon. If the crew ejected, the locator beacon automatically activated to alert rescue forces as to their whereabouts—not good if coming down over unfriendly territory! The next item was to set the interphone to hot mike so the pilot and RSO had instant communications with each other. On all operational missions, the RSO had to select a specific high frequency (HF) radio frequency, listening for important coded information. On a few occasions, the SR-71 had to be called back to the Det because of impending or rapidly deteriorating weather conditions. If other national reconnaissance systems thought the SR-71 was straying off course, they issued a coded HF warning telling the SR-71 crew to recheck their navigation system for accuracy. If the coded message authenticated properly and all the navigational system checked out, the crew pressed on.

Flying in a hostile area, crews had a higher set of aircraft limits, which they could apply in an emergency, called the "tactical limits." If they found the normal operating restrictions unacceptable because of a hostile environment, e,g., a direct threat to the aircraft, the pilot could use the tactical limits spelled out in the SR-71 flight manual. When

Opposite: All the various sensors and their approximate locations when loaded onboard the SR-71. *Lockheed Martin*

148

The dreaded SA-2 "Guideline" missile, a medium- to high-altitude mobile surface-to-air missile system manufactured by the Russians. This is the missile that shot Francis Gary Powers down over the Soviet Union on 1 May 1960. Although over 100 SA-2 missiles have been fired at the SR-71, no planes were ever shot down. *Rich Graham*

authorized by the commander, speeds up to Mach 3.3 could be flown as long as the CIT limit of 427 degrees C (800 degrees F) was not exceeded. Accelerating to Mach 3.3 decreased the margin of flight safety; exposure to these limits had to be as brief as possible. They were to be used only when adherence to normal or emergency restrictions would place the aircraft in a more hazardous situation because of possible hostile actions. By applying the tactical limits, the crew could exit a hostile area at high Mach and altitude, but only for a brief time.

Other than the HF radio crackling in the background, the only communication was between the pilot and RSO. Flying as a formed crew paid off when critical decisions had to be made. Most crews were go-oriented and instinctively knew

neither one would abort a mission unless it was absolutely necessary. There could be no confusion about aborting or pressing on with the mission. The pilot's main job in the sensitive area was flying the aircraft smoothly to get good imagery and watching outside the windows for visual signs of impending threats from aircraft contrails or SAMs being fired. The RSO was closely monitoring the navigational accuracy of the aircraft and looking for any electronic indications in the cockpit that the SR-71 was being acquired by enemy radar.

After exiting the sensitive area, crews reversed the entry checklist procedures and the RSO made a coded "ops normal" call on the HF radio, communicating the mission was successful thus far and everything was going as planned. Around

ON OKINAWA WE FLEW ONE MISSION INFREQUENTLY

that involved gathering intelligence on North Korea. Its profile was unique because it didn't involve a refueling. From takeoff to landing the total mission time was fifty-seven minutes and we named it the "Rocket Ride." On a normal mission the pilot was part systems engineer, inlet specialist, and pilot. This mission was very demanding because everything happened so quickly: departure, acceleration and climb, cruise, and descent for landing. Mentally you had to stay well ahead of the plane on this mission compared to others.

North Korea has long been noted for its ability to hide and be deceptive about its plans and intentions relating to invading South Korea. The demilitarized zone (DMZ) between North and South Korea is 160 miles long and only 2.5 miles wide. The area either side of the DMZ is the most heavily militarized in the world. Over many years North Korea has dug massive tunnels and military storage facilities on the north side of the DMZ in preparation for future hostilities with South Korea.

This was the first "Rocket Ride" for Don and me. Our mission was planned to fly down the middle of the DMZ, imaging directly under the SR-71 and into North Korea about fifty-seven miles. The route of flight took us directly from Okinawa, northward over the Korean Straits, into the Sea of Japan, to make a northeast to southwest pass through the DMZ, then exit into the Yellow Sea and return to Okinawa.

The mission called for a 65,000-pound fuel load, and the takeoff roll on runway 05R was noticeably longer than normal, although not excessive. Still in AB we made a 400-knot climbing left-hand turn to the north and departed Okinawa. Within four to five minutes we were at 33,000 feet and began our supersonic climb and acceleration. Mach 3.0 came quickly, and we leveled off around 73,000 feet. Total time from brake release on Okinawa to Mach 3.0 and 73,000 feet was just short of seventeen minutes. It was extremely busy in both cockpit, but everything was going smoothly . . . so far!

We entered the DMZ and I remained focused on the aircraft's performance and kept a watch on the right side for any signs of hostile action. Don was busy monitoring the sensor systems. He also kept a watchful eye on the radar warning gear telling him we had North Korean surface-to-air missile sites searching or tracking us. Total time over the DMZ would be about seven minutes. Halfway through the DMZ all was calm. Suddenly, I noticed a large object over North Korea that was higher than we were! Whatever it was, we were closing on it fast! I barely had time to tell Don when it rapidly passed directly over and behind us. It turned out to be a high-altitude weather balloon that could have been launched from the USSR, China, North Korea, or South Korea. Don and I agreed later this was probably coincidental, rather than a nefarious plot of some sort.

As we exited the DMZ, our programmed flight track kept the SR-71 heading for China. Our onboard sensors began recording electronic signal intelligence as we flew outside their international airspace. If China's military radar network detected our presence and reacted, the onboard recorders would gather a wealth of intelligence on their capabilities to react to the SR-71's presence. Many of our flights were closely monitored by other U.S. assets. Don always monitored the HF radio for coded transmissions sent from other assets, alerting us to something we possibly might not be aware of.

Shortly after exiting the DMZ, Don received a valid HF radio call. He quickly deciphered the message. We were told to "check our navigation system for accuracy." We had never received a message like this before! After checking our navigation system, we proceeded on track and then began a left hand descent, landing at Kadena Air Base in less than an hour. We found out later that the other assets didn't know our programmed flight path and when they were aware the SR-71 was heading directly for China it raised a red flag. The sheer excitement of this "Rocket Ride" made us look forward to future ones!

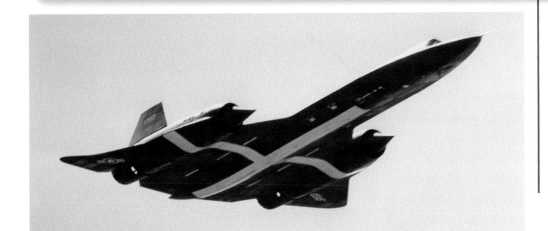

In July 1976 two world speed runs were accomplished in SR-71, tail number 958. The crew of Al Joersz and G. T. Morgan set a speed record over a straight course of 2,193 mph. The crew of Pat Bledsoe and John Fuller set a speed record over a closed course of 2,092 mph. The white cross underneath the plane was used for optical tracking to establish the precise start and stop timing of the speed runs. *Lockheed Martin*

1979, the crew of Tom Alison and J. T. Vida, and Lee Shelton and Barry MacKean, flew a very unique mission. Tom and J. T. were flying from Beale AFB to Kadena. At the right time, the crew of Lee Shelton and Barry MacKean launched out of Okinawa. The mission was to gather intelligence on the Soviet Union's naval fleet headquarters at Vladivostok with both aircraft arriving there simultaneously. With planned altitude separation, both planes produced a head-on Mach 6+ closure for the Soviets to try to figure out. A similar Mach 6+ closure mission was flown by two SR-71s off the coast of Petropavlovsk to gather intelligence on the Soviets' nuclear sub base.

Left to right: crew members Ed Bethart, Maury Rosenburg, Tom Henichek, and Bill Burk pose outside of the Bachelors Officers Quarters 318 that housed all the SR-71 and KC-135 crew members at Kadena AB, Okinawa. *Ed Bethart*

Post-mission picture of two crews just finishing their missions at Det 1, Okinawa, Japan. Standing, left to right: Don Emmons, B. C. Thomas, and Rich Graham. Kneeling, left to right: John Murphy, Jay Reid, and Buz Carpenter. *Rich Graham*

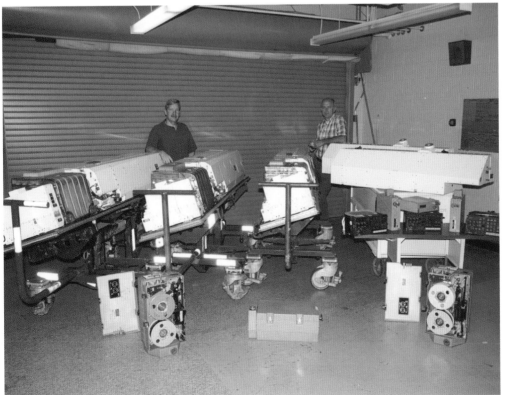

Above: The opaque window in front of the control stick is the map projector. The entire route of flight was loaded on a 35mm film strip. At each turn point, the map displayed the desired Mach number, bank angle, time en route, fuel, altitude, and other important information for the crew to keep up with. The RSO's map display included additional sensor information. *Rick Poole Photography*

Left: Mike Hull (left) and Bill Whittle posing behind the defensive (DEF) systems suites they declassified for museum displays: DEF H, DEF A2C, DEF C, and Universal DEF Control Panel. These units provide electronic warfare jamming. The gray recorder units on the floor recorded electronic intelligence from whoever was trying to locate the SR-71 on radar. Enemy capabilities were obtained by recording their electronic signatures. *Mike Hull*

CHAPTER 18

Descent and "Hot" Tanker Air Rendezvous

The deceleration and descent maneuver left little room for error. Unlike other jets, the pilot didn't have the option to change throttle settings at random or add and subtract drag devices to modify their rate of descent. Once the throttles were brought out of afterburner at Mach 3+ to start down from around 78,000 feet, the bottom-out point of the deceleration and descent was basically set in stone. It was mandatory for the pilot to manage the SR-71's engines and inlets during the descent in a precise configuration to preclude unstarts, compressor stalls, and engine flameouts.

Prior to starting down, the RSO established ranging with the lead tanker, making sure the tankers were there in their orbit. Both crews monitored the radio in case the tanker had something important to report or had to move the ARCP because of bad weather. No news was good news. The RSO referenced his descent chart to include any changes to the planned descent point on his computer generated flight plan. The descent planned for a constant airspeed of 350 KEAS minimum under no wind conditions, standard day temp devs, down to a level off altitude of 25,000 feet. The tankers would be orbiting at 27,000 feet. If any of these factors changed, the SRO needed to make final adjustments to the start descent point. To ensure there was no confusion, the RSO coordinated with the pilot on any variables that he included in his calculations. A typical deceleration and descent from 78,000 feet at Mach 3.0 down to 25,000 feet took around nine minutes and covered approximately 165 NM. The ARCP would then be about 40 NM ahead.

Opposite: On 13 January 1982, the SR-71B model flew its one thousandth sortie. No other Blackbird has flown as many hours or sorties than the trainer 956. At the completion of the record-breaking flight, ceremonies were held on the Beale flight line. The trainer is on display at the Air Zoo Museum in Kalamazoo, Michigan. *Bill Thompson*

Above: The SR-71 on takeoff from Okinawa. In the background, underneath the plane, is the "elephant cage" at Tori Station. The highly classified antenna array received signals from a wide range of the frequency spectrum around the globe, allowing the United States to gather signals intelligence on selected foreign countries.
Tony Landis

Right: The descent profile chart. To determine when to bring the throttle out of afterburner and start down, the RSO had to consider not only the winds, but all of the applicable items in the note section. All of the checklist steps during the descent are shown on the profile chart. The check numbers allowed the RSO to see how well (or badly) the descent was going at various altitudes.
Lockheed Martin

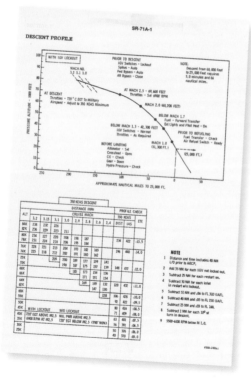

them back further do drop down into the full military power position. As the airspeed began to decrease, the pilot slowly rolled the pitch wheel forward and intercepted a speed between 350–365 KEAS for the descent and deceleration. Since 350 KEAS was the minimum airspeed, 365 KEAS provided an extra margin of safety from unstarts, compressor stall, or engine flameouts. As the airspeed approached 365 KEAS, the KEAS HOLD function on the autopilot was engaged. Just like the climb, the autopilot then controlled the airspeed by varying the pitch of the aircraft.

If an engine were ever to flameout, it would most likely occur during the deceleration and descent. Even holding the parameters perfectly, the engines (one or both) would occasionally compressor stall or flameout for no apparent reason. A few crew members have the dubious distinction of logging the only glider time in the SR-71 by having both engines flameout during the descent. In most instances, successful engine restarts were not obtained at high altitudes and high Mach numbers. The pilot had to be patient, as successful engine starts generally occurred at Mach 1.7 and below.

When the Mach reached 2.5, the pilot slowly retarded both throttles until each revolution per minute (rpm) gauge read 6,900 rpm. Passing through 60,000 feet, the RSO turned on the transponder to enable air traffic control to pick up the SR-71 on radar and display its altitude for separation from other traffic. The crew accomplished a few more checklist items passing Mach 1.7, and upon reaching Mach 1.3, the pilot could finally place the throttle anywhere to adjust the descent profile. Once the plane was subsonic, the pilot maintained Mach 0.9 for cruise, disengaged the KEAS HOLD function on the autopilot, and leveled off from the descent. At this point the SR-71 should be around 40 NM from the ARCP and 1,000 feet below the refueling altitude. From here, the flying merely repeated the cold rendezvous procedures.

Not every rendezvous went smoothly. On a hot rendezvous, if the SR-71 crew

After discussing the agreed–upon start descent point, the crew members double checked that each of their cockpits displayed the same distance to that point. Arriving at the computed start descent point, the pilot slowly retarded both throttles to minimum AB, paused there momentarily, then pulled

initiated its supersonic descent too late, or the tanker crew made its final turn too early, it was possible to end up with what was called a "tanker overrun." In this scenario, it was highly probable that the SR–71 would end up in front of the tankers rather than behind them. When a potential overrun was called out over the radio, by either the tanker or SR–71 crew, the procedure was for the tanker to descend and accelerate immediately and the SR–71 pilot to slow down as quickly as possible. The SR–71 pilot could often do S-turns to help increase the distance between the two aircraft. Although sometimes exciting, the rendezvous eventually sorted itself out. Once the plane was refueled, the process of climb, cruise, and descent would be repeated again in another other part of the world.

Above: The SR-71B trainer after being refueled before starting out on another Mach 3+ leg. With the raised rear cockpit, the vertical distance from the boom to the rear cockpit has a narrower tolerance than the SR-71A. *Lockheed Martin*

Left: Pilot B. C. Thomas gets hosed down after completing his last flight in the SR-71. B. C. ended up with 1,217.3 hours in the Blackbird, one of nine crew members achieving over 1,000 hours. Coming down the steps is his RSO at the time, J. T. Vida. *B. C. Thomas*

CHAPTER 19

The Recovery

Opposite: Landing on runway 05R at Kadena AB, Okinawa, Japan. Flying final approach at 175 knots minimum combined with the relatively nose high attitude, the pilot had little room for error. *Don Emmons*

Below: An SR-71 on a left base leg to final approach for runway 15 at Beale. The gear is down and locked and the base leg is a descending turn at 230 knots. The SR-71 has no flaps, slats, speed brakes, or any other means to fly slower. *Doug Soifer*

After one or numerous other hot legs and refuelings, it was finally time to land. Thirty to forty-five minutes prior to the planned arrival, the tower officer reported back to the tower for the landing and the mobile crew stopped by operations to pick up the Det commander. It was rare for the commander to miss the landing and recovery of the SR-71. The mobile crew positioned the car on a taxiway that allowed it to accelerate rapidly, eventually catching up with the SR-71 during its landing roll.

For fuel economy, the final descent began as close to the home base as possible. Since Okinawa was a tiny island surrounded by the Pacific Ocean, the aircraft could be supersonic until the last possible minute—one continuous descent from Mach 3+ cruise to landing. At Mildenhall, the SR-71 had to be subsonic prior to crossing any landmass in the United Kingdom. The crew could use radio silent procedures all the way to a landing, but normally established radio contact, requesting the type of approach they wanted. Descending through 60,000 feet, the crew made radio contact with the appropriate controlling air traffic agency. Once subsonic, around 30,000 feet, the pilot began using the pitot static instruments again and resorted to hand flying, if he wasn't already.

The crew made a courtesy call to the mobile crew over the UHF radio about thirty minutes out from landing, advising them of anything unusual. Many operational missions were time sensitive and had a high priority on getting the intelligence out right away, requiring the pilot to land as quickly as possible. If not, pilots often begged for a touch-and-go landing to get as much practice in as they could.

Final approach and landing speeds were based on the weight of the plane. The basic final approach speed was 175 knots (200 mph). Touchdown speed was 155 KIAS (178 mph) with 10,000 pounds of fuel remaining. For each additional 1,000 pounds of fuel, crews added one knot extra to both the approach

and touchdown speeds. Typically, the pilot lowered the landing gear between five to ten miles out from the runway. After lowering the landing gear, the pilot took one last look at his annunciator panel, looking for indications of anything abnormal with the aircraft. Now he devoted full attention to the landing.

On a straight-in approach, it took very little power to hold 175 KIAS. The final approach pitch attitude was relatively high—somewhere around ten degrees nose up. Forward visibility during landing was good, although the long nose blocked out runway references beneath and to the sides of the aircraft. As the pilot neared the runway overrun, he slowly retarded the power to idle and decreased the airspeed. Normally, the SR-71 was flown directly to touchdown rather than attempting to hold it off the runway. The massive delta wing of the Blackbird produced a large amount of ground effect, allowing the SR-71 to float, cushioning the landing for

a smooth touchdown at 155 KIAS. If a go-around was necessary, the pilot accelerated the engines rapidly moving the throttles from idle to either full military power or afterburner power.

As soon as the main gear touched down, the pilot's first action was to reach forward with his left hand and pull out the drag chute T-handle, deploying three chutes in succession. The first chute was a forty-two-inch pilot chute, followed by a ten-foot extraction chute, and finally the forty-foot drag chute. The pilot chute and extraction chute were not attached to the aircraft and fell onto the runway. The pilot maintained the nose-high landing attitude for aerodynamic braking. Once the main chute fully blossomed, he slowly lowered the nose wheel onto the runway by relaxing back pressure on the stick. The nose wheel touched down around 100 KIAS, as the SR-71 slowed down rapidly once the drag chute took full effect. The pilot checked his brakes gently to ensure they were working

SR-71 964 on short final landing on runway 15 at Beale AFB. The aircraft is directly over the 1,000-foot overrun that can be used in emergencies. The mobile car is already accelerating to catch up.
Lockheed Martin

Left: The massive forty-foot drag chute fully deployed. It gives a good tug on the crew member's shoulder harness once it's fully blossomed and slows the plane down rapidly. Very little braking is required until the chute is jettisoned on the runway. *Lockheed Martin*

Below left: Landing roll out on runway 11 at Det 4, RAF Mildenhall, England. The large hangar in the background was the operations facility used for mission planning by the crews. The Det 4 commander, operations officer, and mission planners also had their offices in the building. *Lockheed Martin*

satisfactorily, and then, at 55 KIAS, he pushed in the drag chute T-handle jettisoning the main chute onto the runway. A maintenance truck followed the aircraft down the runway to pick up the three chutes, allowing other aircraft to land.

Just before the plane rolled past the mobile crew, the driver floored the car onto the runway, chasing after the aircraft. Racing down the runway, the mobile crew watched for blown tires and full drag chute deployment. If anything appeared abnormal during the landing rollout, the mobile crew advised the flyers over the radio. The mobile car caught up with the aircraft about the time the drag chute was jettisoned onto the runway. It passed alongside the SR-71 with a thumbs-up to the flyers, then led the aircraft back to the hangar.

Once the tower officer observed the SR-71 was at a slow taxi speed, he instructed tower personnel to disarm the barrier. After the mission monitor finished making all the classified notifications and filling

Above right: To stay cool in the cockpits, crews kept the canopies down during the taxi. About now, maintenance personnel drove their truck down the runway to pick up the forty-two-inch pilot chute, the ten-foot extraction chute, and the forty-foot drag chute lying on the runway. *Lockheed Martin*

Above: Taxiing in the hangar at Beale AFB after a night mission. As soon as the post-flight checks are accomplished and the engines shut down, maintenance personnel swarm all over the plane with specific tasks to accomplish. Downloading the sensors had top priority. *Lockheed Martin*

Right: To cool down the brakes after taxiing in from landing, portable cooling fans were placed behind each main landing gear. With a light fuel load after landing, the aircraft wanted to pick up speed and the pilot had to be careful not to overheat the brakes. The tires were rated for a maximum speed of 239 knots. The 32-ply tires were silver coated to reflect heat and were good for about fifteen to twenty landings before changing. *Rich Graham*

out various reports, he drove back to operations. Some of the first tanker crews might have been on the ground already, busy debriefing with tanker operations how the mission went from their viewpoint and what they could do to make it better next time. The aircraft's crew chief had been making arrival preparations in the hangar, while tech reps and other maintenance personnel started to gather, awaiting the SR-71 return.

For all operational sorties, the SR-71 taxied directly inside the hangar for maintenance to expedite downloading of the sensors. The mobile car preceded the plane to the hangar entrance and parked to the side while the aircraft slowly entered the hangar. The pilot followed the crew chief's hand signals, guiding the plane safely into the tight confines of the hangar, until the signal was given to stop. After the crew chief gave the "chocks installed" hand signal, the pilot released the brakes. The crew chief plugged his headset into the intercom system and checked in with the crew for the remainder of the shutdown. After completing numerous steps of the checklist confirming redundancy of various systems, the pilot placed both throttles in the cutoff position.

After engine shutdown, each crew member unlocked and opened their canopy. After breathing 100 percent oxygen for the entire flight, it was good to feel fresh air again. The first personnel to greet the crew at the top of the steps were the PSD technicians. Before attempting to climb out of "hot" ejection seats, the crew waited patiently for PSD technicians to insert the ejection seat and canopy safety pins, disarming the seats. After the seats were

safely pinned, the crew members unlocked their pressure suit helmets and handed them to PSD, followed by the gloves, checklist, and other mission materials. Once PSD disconnected the parachute harness, hap belt, survival kit straps, stirrups, oxygen hoses, parachute D-ring, communications cord, face heat cord, and cooling air supply hoses, he was free to stand up and exit the cockpit safely.

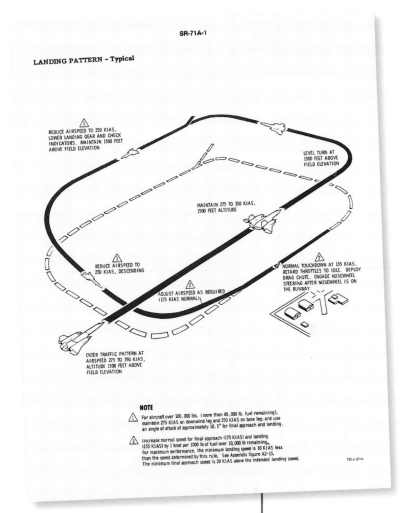

SR-71A-1

LANDING PATTERN - Typical

REDUCE AIRSPEED TO 250 KIAS. LOWER LANDING GEAR AND CHECK INDICATORS. MAINTAIN 1500 FEET ABOVE FIELD ELEVATION

LEVEL TURN AT 1500 FEET ABOVE FIELD ELEVATION

MAINTAIN 275 TO 350 KIAS. 1500 FEET ALTITUDE

REDUCE AIRSPEED TO 250 KIAS, DESCENDING

NORMAL TOUCHDOWN AT 155 KIAS. RETARD THROTTLES TO IDLE. DEPLOY DRAG CHUTE. ENGAGE NOSEWHEEL STEERING AFTER NOSEWHEEL IS ON THE RUNWAY

ADJUST AIRSPEED AS REQUIRED (175 KIAS NORMAL)

ENTER TRAFFIC PATTERN AT AIRSPEED 275 TO 350 KIAS, ALTITUDE 1500 FEET ABOVE FIELD ELEVATION

NOTE

For aircraft over 100,000 lbs. (more than 40,000 lb. fuel remaining), maintain 275 KIAS on downwind leg and 250 KIAS on base leg; and use an angle of attack of approximately 10.5° for final approach and landing.

Increase normal speed for final approach (175 KIAS) and landing (155 KIAS) by 1 knot per 1000 lb of fuel over 10,000 lb remaining. For maximum performance, the minimum landing speed is 10 KIAS less than the speed determined by this rule. See Appendix figure A2-15. The minimum final approach speed is 20 KIAS above the intended landing speed.

This is a the normal overhead pattern flown by the SR-71. The fuel required for a standard instrument pattern is 3,000 pounds and for a closed pattern after a go-around requires around 1,000 pounds of fuel. *Lockheed Martin*

CHAPTER 20

Post-Flight Activity

As soon as the last engine was shut down, maintenance personnel immediately swarmed all over the aircraft, like worker bees in a hive, each with a specific job. It was a sight to behold—everyone working on the aircraft in perfect harmony. The highest priority was to download the sensors from their bays underneath the fuselage. At the foot of the ladder, waiting for the crew to step down, were the Det commander, maintenance officer, mobile crew, and the aircraft's crew chief. Everyone was interested to hear about the plane and the mission. It was a quick debriefing at the aircraft, lasting only five to ten minutes.

The crew then climbed into the PSD van for the drive back to de-suit. During the drive, their stirrups were removed and boots unzipped. PSD personnel recorded several routine questions about the flight and asked if the crew had any problems with the suits. Once inside PSD, it only took a few minutes to remove the boots, peel off the pressure suit and sweaty long-john underwear, then head for a shower. The crews changed back into their flight suits, gathered up their classified mission materials, and were off to start the formal debriefing process.

Operations and intelligence debriefings came first. Mission planners wanted to know if there were any unusual sightings, any problems flying the route, air refueling problems, sensor operations, and the state of the target area weather. If hostile actions were taken against the SR-71, their debriefing could last considerably longer than the normal ten minutes or so. Weather personnel were especially interested in temp devs during the climb and cruise, turbulence, wind speed, and any other significant weather conditions. The operations officer wanted to know how it went from his vantage point in the tower. The mobile crew asked if there was anything they could have done better to make the operations run smoother. The entire debriefing process took around fifteen to twenty minutes if nothing unusual happened. The flyers and mobile crew then stored their classified materials in the safes.

The next stop was the maintenance debriefing. Fifteen to twenty maintenance personnel sat around a large table, representing the major aircraft systems (hydraulics, electrical, flight control, engine, inlets, avionics, sensors, navigation, fuel, etc.). The RSO had already filled in

Opposite: The crew of Terry Pappas and John Manzi congratulating each other at the top of the stairs. PSD technicians took the helmet and gloves so crews could get some ventilation. *Lockheed Martin*

Right: Now that's a good mobile crew, handing Tom Danielson a cold beer after a long flight. Crews were allowed to take their drink of choice (within reason) on flights to stay hydrated. It was put into a clear, plastic, squeeze bottle with a long drinking straw attached. At the bottom, right side of the pressure suit helmet was the feeding port for the straw to be inserted and then to the mouth. *Lockheed Martin*

Below left: As soon as the sensors were downloaded from the plane and onto their dollies, they were immediately taken to processing vans where the intelligence collected was turned into meaningful information. Intelligence personnel worked long and late hours to get the information disseminated to the users. *Lockheed Martin*

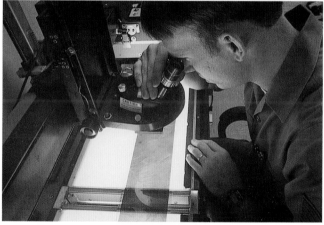

Above right: A technician looks through the lens to evaluate the quality of the pictures taken by the SR-71. One of the post mission activities was analyzing the photographic intelligence as soon as possible. *USAF*

a pre-formatted maintenance form and handed it to the maintenance supervisor conducting the debriefing. The supervisor began by reading aloud all that information (time above certain Mach numbers, maximum Mach and altitude, air refueling offloads, aircraft discrepancies, etc.). Each representative recorded the information, then asked the crew specific questions about their particular aircraft system. Once the formal maintenance debriefing was concluded, the crew remained behind to discuss any aircraft maintenance problems in greater detail. Between a thorough debriefing from the crew and the information recorded inflight by the DAFICS, maintenance was able to troubleshoot and correct discrepancies accurately.

After the maintenance debriefing, the crew drove back to their bachelor officer quarters or the officers' club for some well-deserved liquid refreshment. Sometimes, either the mobile crew or the third crew would host a so-called "debriefing" back

Above: The last SR-71 loss occurred on 21 April 1989 in aircraft number 974. A double hydraulic leak forced Dan House and Blair Bozek to safely eject into the Philippines Sea, a quarter-mile north of Luzon. The decision was made to recover the aircraft from the bottom of the sea. *Don Emmons*

Left: Representative Larry Hopkins being congratulated by me after his DV flight in the SR-71B in 1988. Behind him is Rod Dyckman, the instructor pilot for the Mach 3+ flight. Congressman Hopkins served on the Armed Services Committee. All DV flights had to be approved by the Pentagon. *Lockheed Martin*

in their BOQ room. Everyone was able to relax, have a vodka and tonic (Habus named this drink a "hook"), discuss the mission, and talk about how well or how poorly the plane performed. The mission crew could now make plans for the next two or three days off. The mobile crew would be flying the next scheduled mission, and their mobile crew would be the pilot and RSO just coming from a few days off.

Often, the crew that was not flying or on mobile duty had projects to do at the Det, tours to give, or personal things to take care of. Occasionally, there were two SR-71s scheduled to fly a mission, requiring that everyone be involved. After each

crew's six-week tour of duty at Det 1 or Det 4, the KC-135Q redeployer flew them back to Beale along with other personnel and maintenance supplies. Back at Beale, they took a few weeks off, flew the SR-71 and T-38, and spent considerable time in the simulator brushing up on emergency procedures. Once Mildenhall became a full-time operation in the late 1970s, crews spent six weeks at Beale before heading off to Mildenhall. After six weeks of flying at Mildenhall, it was back to Beale for another six weeks, then off to Okinawa for six weeks. Sometimes crews spent well over half a year away from Beale—long separations, but the flying was great!

Above: SR-71 with a collapsed left main landing gear in the mid-1980s. Maintenance was repacking the landing gear strut seal, which required them to do an emergency gear extension while on jacks. They put gear safety pins in the right and nose gear, but forgot the left gear. When another maintenance crew arrived and applied hydraulic power with the gear handle still in the up position, the left main gear collapsed.
Tony Landis

Opposite: SR-71 955 flying in formation with a P-51 Mustang: the world's fastest propeller and jet aircraft. Notice the different pitch attitudes of each plane. SR-71 is flying as slow as possible with the nose up high and the P-51 is level.
Bill Flanagan

CHAPTER 21

The Final Years

Highly classified programs within the Department of Defense (DOD) are financed under a special funding category called the "black world." The objective is to protect black world money from scrutiny and provide programs with relatively stabilized funding levels so that they do not have to struggle or compete with the rest of the DOD budget. Black world funding also permits a high-risk, technological advanced program to emerge from a research and testing phase into the development and prototype phase stage without the problems and oversight that plague the DOD "white world" programs.

Originally, the SR-71 program's funding was contained within the black world. In the mid-1970s, SR-71 funding was moved from the black into the white world. For the first time, the SR-71 program had to compete for its funding along with the rest of the DOD budget. Through every annual budget cycle an office in the Pentagon, called XOORZ, became the only advocate and spokesperson for funding its annual budget. The office had to justify, defend, and argue the SR-71's requirements against the needs of the Air Force. Each budget cycle was an annual frenzy, full of sharks and pitfalls.

The SR-71 was under the command of SAC. Their mainstream, and logically their highest funding priorities, were two legs of the nuclear triad: nuclear intercontinental ballistic missiles and nuclear equipped B-52 bombers. SAC's ICBMs and B-52s were in need of expensive upgrade. New ICBMs and the B-1 bomber were also trying to make their entry into SAC's inventory during the early 1980s. They received high priority on SAC's funding list, and it became obvious SAC's interest in the SR-71 was slowly waning.

The intelligence community, believing overhead satellites were the way of the future, became more vocal in comparing the SR-71's high operating costs to that of satellites. During my four years in the Pentagon (1982–1986), I became increasingly aware that defending and justifying the SR-71 budget was becoming difficult. Support was slowly disappearing. It became a duel which the SR-71 program was slowly losing.

Although government agencies, such as the CIA and NSA, were slowly abandoning their support for the SR-71 program in favor of satellites, the synoptic coverage (displaying conditions as they exist simultaneously over a broad area) provided by the SR-71 was far superior to satellite reconnaissance. Broad-area coverage from different approach angles, in a relatively short time span, produced

considerably better intelligence than a predictable, single satellite pass, every ninety minutes. In 1990, the SR-71 was the only airborne reconnaissance platform that could penetrate hostile territory, accomplish wide-area synoptic coverage, and still survive. It could also be tasked reasonably quickly and had the element of surprise.

The SR-71 program was originally scheduled for closure in 1989. However, with last-ditch efforts from Blackbird supporters in Congress, and a few in the Pentagon, it clung to life for another year. One SR-71 pilot who worked in XOORZ, Lee Shelton, had often stated that we were our own worst enemy. He observed from working in the Pentagon that the Blackbird program had such high secrecy from others, when the time came to look for supporters and proponents, they were scarce. Very few individuals of influence had any inkling of our capabilities. It was extremely difficult to find support in Congress and in the Pentagon if your reconnaissance capabilities were not fully understood.

I believe the SR-71 program would have been better served if it had been under the control of an intelligence agency in Washington. At the very least, an intelligence agency would have been more receptive than SAC to upgrading the SR-71s with state-of-the-art intelligence sensors, enabling them to compete well into the twenty-first century. Others will argue, but I believe the SR-71 never had a legitimate place in SAC for the long term.

In one respect, SAC's lack of support for the SR-71 program was understandable. They had to pay all the bills, but derived very little payback. The vast majority of the SR-71's intelligence data was primarily used by other government agencies and theater commanders in Europe and the Pacific, rather than SAC. Consequently, from 1986 to 1990, what little support there was for the SR-71 program at SAC headquarters turned negative. During this same time frame, the chief of staff of the Air Force, Gen. Larry D. Welch, was not an advocate of the SR-71. Basically, if it didn't carry bombs or shoot missiles, he didn't want to hear of it. With the lack of support for the SR-71 program at such high levels, the end was approaching.

One of the more subtle (albeit, highly important) reasons the SR-71 program was increasingly under the gun was its decreasing general officer sponsorship. The first seven 9th Wing commanders (January 1966 to September 1977) were all SR-71 pilots and continued in their Air Force careers to become general officers. As they moved to positions of greater responsibility within the Air Force, they still supported the SR-71 program because they understood its capabilities. Just like the "fighter mafia" within the Air Force that promotes new and

In February 2006, crew members, maintenance personnel, and technical representatives put on a two-day symposium at the Pima Air & Space Museum in Tucson, Arizona. Gracious hosts Don and Mary Stein opened their house for a dinner party where we could all get together with our wives and relive the days when Blackbirds ruled the skies. *Mary K. Stein*

better-manned fighter aircraft, the SR-71 had its sponsors and supporters early on. Former SR-71 pilots, including Generals Bill Campbell, Doug Nelson, Charlie Minter, Bill Hayes, Pat Halloran, John Storrie, Mel Vojvodich, and Colonels Willie Lawson and Ken Collins, retired during the 1980s and left a major void in SR-71 support throughout the Air Force. The most noted advocate of all was Gen. Jerome "Jerry" O'Malley, who flew the very first SR-71 operational sortie in March of 1968. As the commander of Tactical Air Command, he died in an aircraft accident with his wife while heading to a speaking engagement in Pennsylvania in April of 1986.

I'm not suggesting all these former Habus were actively campaigning on behalf of the SR–71, but can state with reasonable assurance that other general officers who wanted to see the SR-71 program cancelled remained silent as long as the former Habus were in a position to use their knowledge and influence within the Air Force. By 1986, practically all the SR-71 general

officers had retired and the program was left with very little internal support. It was just a matter of time before a new generation of generals would scrutinize the SR-71 program closely, looking for a means to fund their own pet projects.

Over the years, our customers were also our biggest advocates. These included the CIA, Defense Intelligence Agency (DIA), National Security Agency (NSA), Foreign Technology Division (FTD), theater commanders at Pacific Air Force (PACAF) and at United States Air Force in Europe (USAFE), and several foreign countries. Many of these agencies were instrumental in keeping the program strong during its early years but later favored satellites and other technical means of gathering intelligence.

Surprisingly, during the late 1980s, the U.S. Navy was the SR-71's greatest advocate. Satellites could not match the SR-71's ability to locate nuclear submarines for the Navy in certain regions of Europe. The Navy takes great pride in being able to constantly track every sea-going combatant

Flight line at Beale AFB after the retirement of the SR-71 was announced. Maintenance personnel moved all eleven SR-71s out of their hangars and took this final picture in honor of the Blackbird program. *Lockheed Martin*

SR-71 flying over one of the many lakes in the Sierra Nevada mountain range near Beale AFB. *Lockheed Martin*

vessel on the globe. Whenever the Navy lost track of a Soviet submarine carrying nuclear ballistic missiles, it called on the SR-71 to fly over the Barents and Baltic Seas, hoping to locate the missing sub. Each branch of the military services is so protective of its individual roles and missions, it seems ironic that the SR-71 program's chief advocate became the U.S. Navy.

Between 1987 and 1988, two former Habus working in XOORZ developed numerous funding proposals for both SAC and the Air Force, showing various plans to retain the SR-71 in the inventory. It was a gamble, but if they could sway the senior leadership that a reduced number of SR-71s, flying at a lower sortie rate, could cover the globe at a cheaper cost, they might be able to fend off those who wanted the program completely terminated. One of the proposals had two aircraft stationed at Det 1 and Det 4, and three at Beale, including the trainer model, for an annual cost of $155 million.

Another proposal had five SR-71s flying at Beale and the two Dets closed down to caretaker status. SR-71s would deploy from Beale to a forward-operating location and fly sorties when hot spots developed around the world. General Welch wouldn't hear of it and didn't want anything short of complete termination of the program. Around the same time, SAC headquarters came out with its operational plan for the shutdown of the SR-71s at both Dets and at Beale. The plan had lots of guidance, but offered little help in execution. Throughout 1987 and 1988, emotions were running extremely high, as no one wanted to move a muscle to help terminate the program. Weekly, there were new rumors that the program might still survive in some capacity. The last thing anyone wanted to do was to go too far in closing the program, preventing it from being revived. Everyone knew that at some point during the program's termination it would eventually reach that point of no return.

In the end, funds never materialized to continue the SR-71 program. At Beale, Det 1 and Det 4 talk began to revolve around

Above: The SR-71B trainer aircraft flying in refueling formation with a KC-135 tanker at low level. NASA was participating in the world's largest air show at Oshkosh, Wisconsin. Having the SR-71 appear at any air show always drew a crowd. *NASA*

Left: Don Emmons addressing the crowd at Plant 42, Palmdale, California. With help from Lockheed Martin, Don arranged to have a final gathering of Blackbird supporters after the closure of the program in 1990. At the time, no one had any idea the SR-71s would return in 1995. *Lockheed Martin*

Ed Yeilding making a final adjustment to his silk gloves. Using the silk inserts made it much easier for the pressure suit glove to slide on and lock in place at the wrist. *Lockheed Martin*

assignments for personnel and what to do with the aircraft's assets and facilities. Confusion reigned. Practically every day some officer at SAC headquarters phoned to ask me (as the 9th Wing commander) why certain steps of the closure were not being taken at Beale or the Dets.

Even as the program was approaching its scheduled closing day of 30 September 1989, there was still hope among ardent SR-71 supporters that the program would remain alive in some fashion. During the summer of 1989, the SR-71 participated in the international Paris Air Show and the Experimental Aircraft Association air show in Oshkosh, Wisconsin. While waiting a final decision, crews continued to fly proficiency sorties between 1 October

1989 and 22 January 1990, and operational sorties right up to the bitter end, the last one occurring on 7 November 1989. It was flown twenty years, seven months, and seventeen days after the first SR-71 operational sortie, and just two days before the Berlin Wall began to crumble.

After months of turmoil, filled with constant rumors and agonizing speculation, SAC established a date for the final SR-71 flight: 26 January 1990. It was only fitting that the last 1st SRS commander, Lt. Col. Rod Dyckman, and his RSO, Lt. Col. Tom Bergam, flew aircraft 960 on the last SR-71 sortie. Rod flew beautiful passes over the field that day at Beale. Although retired from the Air Force at the time, I was there for the highly emotional and somber event. I found it hard to believe that the skies over Beale would never again be graced by its presence. For so many years, the SR-71s were part of Beale's landscape, and suddenly it felt unreal that this was to be the last flight.

On 6 March 1990, nearly two months after the SR-71 was officially retired from the Air Force, an SR-71 (972) set four international speed records while being delivered to the Smithsonian National Air and Space Museum at Dulles International Airport, Washington, D.C. The aircraft was

After Rod Dyckman and Tom Bergam flew the final SR-71 flight on 26 January 1990 at Beale AFB, a plane-side ceremony retires the colors for the last time. Pilot Rod Dyckman and Ben Rich, co-designer of the Blackbird family of aircraft and president of Lockheed's famed Skunk Works, sadly watch two Air Force sergeants fold the Blackbird flag for the last time. *Lockheed Martin*

JT Vida and I felt extremely fortunate to fly the SR-71 Blackbird on its record flight to the Smithsonian Institution. Fantastic aircraft! And JT was an outstanding aviator, liked and admired by all. We miss him tremendously.

Sincerely,
Ed Yeilding
Pilot
"Coast-to-Coast"
Aircraft Speed Record
67 minutes 54 seconds
March 6, 1990

flown by pilot Lt. Col. Ed Yeilding and RSO Lt. Col. Joe "J. T." Vida, who were assigned to the Palmdale, Plant 42, Flight Test Facility. The record-setting flight covered 1,998 NM in 68 minutes, at a speed of 2,176 mph. The aircraft is now on display at the Smithsonian's Udvar-Hazy Center at Dulles airport.

Many in Congress also knew the true capabilities of the SR-71. The day after the record flight, Senator John Glenn summarized the premature retirement of the SR-71 when he spoke before Congress on 7 March 1990 and stated for the congressional record,

In view of the high costs of other Air Force programs, the costs of this program and its benefits were both affordable

and reasonable. The SR-71 provides coverage on demand with little or no warning to the reconnaissance target—it is a highly flexible system. The SR-71 is able to penetrate hostile territory with comparatively little vulnerability to attack, unlike other reconnaissance platforms. While opponents of the SR-71 have argued that national technical means are capable of performing the same mission, these systems are far less flexible and survivable than the SR-71. In retiring the SR-71, the United States has essentially removed itself from the strategic aerial reconnaissance business. Intelligence systems such as the SR-71 are the eyes and ears for our nation's defense and are therefore true force multipliers. Mr. President, the termination of the SR-71 program was a grave mistake and could

J. T. Vida (left) and Ed Yeilding pose prior to the record-setting flight from Oxnard, California, to Washington, D.C. Sadly, J. T. passed away in September 1992. His love and passion of flying the SR-71 is exemplified by the fact he is our all-time highest flying crew member with 1,392.7 hours and seventeen years of continuously flying the Blackbirds. *Lockheed Martin*

Above left: Ben Rich gives a thumbs up to the record breaking crew of Ed Yeilding (left) and J. T. Vida for a job well done. Ben Rich took over the Skunk Works after Kelly Johnson in 1975 and remained the boss until retirement in 1991. He was considered the father of stealth technology. *Tony Landis*

Above right: The Air Force wanted very little to do with the publicity surrounding the SR-71 speed record. Consequently, few dignitaries from the Pentagon were there for the ceremonies. On hand were two former SR-71 crew members, also SR-71 speed record holders, Col. Al Joersz (left) and Brig. Gen. Buck Adams to greet the arriving record breakers. *Tony Landis*

Right: Ed Yeilding (left) and J. T. Vida receiving their National Aeronautical Association plaques for setting the world record speed flight. *Lockheed Martin*

place our nation at a serious disadvantage in the event of a future crisis.

When the program finally closed down in 1990, the National Air & Space Administration (NASA) saw an excellent opportunity. They were looking for a high-speed aircraft to use in testing future supersonic and hypersonic engines and aircraft. NASA borrowed two SR-71A models, the SR-71B model trainer, and the SR-71 simulator, which was shipped to their facility at Edwards AFB, California. With their small fleet of three Blackbirds, it meant that sole-source vendors, who produced highly specialized items such as the oil, hydraulic fluid, and JP-7 fuel, would still be in operation.

Above: NASA's linear aerospike SR-71 experiment, LASRE, used the SR-71 at Edwards AFB as a "flying wind tunnel" to test the aerodynamics of a future reusable launch vehicle. Here water is being dumped after the first inflight cold flow test with gaseous helium and liquid nitrogen on 4 March 1998. *NASA*

Left top: NASA pilot Steve Ishmael (right) and flight engineer Marta Bohn-Meyer meet the Lockheed SR-71 program manager, Jay Murphy, after flying from Edwards AFB to Palmdale on 12 January 1995. Marta was the only female crew member in the history of the SR-71. An accomplished aerobatic pilot, serving twice on the U.S. aerobatic team, Marta died while practicing for the 2005 championships. *Lockheed Martin*

Left bottom: The unique display of 964 at the Strategic Air and Space Museum just outside of Omaha, Nebraska. Mounted on pillars, visitors are able to view the SR-71 from the lower level. The nose section is obvious by the darker color and the air refueling door is in the open position. *Doug Soifer*

CHAPTER 22

Rising from the Ashes

Former senators Glenn, Byrd, Stevens, and Nunn, who sat on influential intelligence committees in Congress, realized there were shortfalls in the United States' ability to gather timely intelligence from anywhere around the world. In 1995 Congress directed the Air Force to reactivate three SR-71s, and appropriated $100 million to do so. To the Air Force leadership, seeing congressional language in a bill to bring back retired Blackbirds was like having their worst nightmare come true.

As it turned out, the Air Force considered the B-model trainer as one of the three aircraft; therefore only two SR-71s were brought back into service. Don Emmons, who had the unpleasant job of dismantling the Blackbird fleet in 1990, was hired as a contractor to help put it all back together again. He was the only person who knew where everything was disposed of in 1990. Three former pilots and RSOs volunteered to leave their current Air Force jobs and returned to fly the SR-71 again. It was decided that since NASA already had the SR-71 simulator at Edwards, and the basic infrastructure to support and maintain its SR-71 operation, Edwards AFB (rather than Beale) would be the best location for the Air Force to re-train and fly its SR-71s. The unit was called Det 2 of the 9th Reconnaissance Wing at Beale AFB.

Using Lockheed maintenance, civilian contractors, and Air Force operations personnel, Det 2 was soon fully operational, ready for deployment anywhere around the globe. Everyone knew instinctively that the two SR-71s had to be upgraded if they were going to compete with other reconnaissance assets. They modified their two SR-71s with a common data link, providing near real-time imagery with a 300 NM line-of-sight. An electrical optical (EO) capability was being integrated into the cameras as well. Two successful flights had been accomplished at 79,000 feet, giving near real-time EO imagery. Many sophisticated improvements were made to ground equipment, making it portable and compatible with existing military processing capabilities, and permitting immediate worldwide transmission of intelligence.

Opposite: When the SR-71s retired in 1990, all the crew members went on to other Air Force assignments. These six SR-71 crew members volunteered and were selected to return to fly the SR-71 at Edwards AFB, at the newly formed Det 2. (L to R) Jim Greenwood, Don Watkins, Mike Finan, Gil Luloff, Blair Bozek, and Tom McCleary. *Tony Landis*

Above: The Air Force transfer ceremony for the SR-71 reactivation at Palmdale, California, on 28 July 1995. Standing (left to right) are Air Force members Terry Melanson, Chris Bennett, Jim Greenwood, Don Watkins, Blair Bozek, Tom McCleary, Mike Finan, and Gil Luloff. At the podium is Lockheed Skunks Works representative Jay Murphy ceremoniously turning over the plane to the Beale 9th Wing commander, Gen. Bill Rutledge. *Tony Landis*

Right: Tail number 962 on display in the American Museum at Duxford Imperial War Museum, England. This aircraft flew the first "completed" operational mission in Europe on 7 September 1976, flying out of RAF Mildenhall. *Martin Lupton*

Despite the wishes of Congress, the Air Force had the upper hand on whether the Blackbirds would ever fly operationally or not, and they never did. In 1997, the Air Force finally found a way to get rid of its nightmare when the White House was looking at the Pentagon for items to eliminate from the budget. The SR-71 program was offered up by the Air Force, and in October 1997, President Clinton line-item vetoed the entire Air Force SR-71 program. From that day on, all Air Force flights in the SR-71s ceased. During the reactivation, Det 2 flew 150 training sorties

and 365.7 hours with its three aircraft without an incident. Once again Don Emmons was put in charge of dismantling the SR-71s, and in October of 1999, he was the last man to walk out the door of Det 2.

NASA continued to fly their SR-71s, with the final flight of any SR-71 occurring on 9 October 1999 at Edwards AFB during their annual open house, flown by the NASA crew of Rogers Smith and Robert Meyer at a speed of Mach 3.21 and an altitude of 80,100 feet.

Over the years many flying records were set by the SR-71. The following is a list

Picture of A-12 tail number 931 being swallowed up by a C-5 for shipment to the Minnesota Air National Guard Museum in St. Paul, Minnesota. It was eventually sent to the CIA headquarters where it is now proudly displayed as a national treasure. *Lockheed Martin*

Former crew members, maintenance, and Lockheed and civilian support personnel meet to put on an SR-71 symposium, sharing stories, tales and legends of flying the Blackbird with aviation enthusiasts. On 2 February 2006, at the Pima Air & Space Museum in Tucson, Arizona, (L to R) Dave Burns, Tom Keck, Bob Gilliland, Ed Yeilding, and Barry MacKean sign posters for the museum. *Rich Graham*

of lifetime achievements by the Blackbird family of planes (A-12, FY-12, and SR-71):

Total Sorties: 17,300, operational 3,551
Total Hours: 53,490 hours total,
    11,008 hours operational
Mach 3+Time: 11,675 hours,
    2,752 hours operational
Crew members over 300 hours: 163
Crew members over 600 hours: 69
Crew members over 900 hours: 18
Crew members over 1,000 hours: 8
Total A-12 pilots who flew operational
    reconnaissance missions: 6
Total SR-71 pilots who flew operational
    reconnaissance missions: 86
Total SR-71 RSOs who flew operational
    reconnaissance missions: 86
Most SR-71 flying time:
    Lt. Col. J. T. Vida with 1,392.7 hours

It's difficult for those who didn't belong to imagine the closeness of this small group of aviators and associates. Those of you who belonged to a fraternity or sorority may be able to partially understand this special relationship. And those of you who have shared a combat experience, i.e., lived and died together, can understand the higher level of brotherhood reached among Habus. As one who has experienced combat, I must tell you that there is no higher level of comradeship, brotherhood, or friendship than that experienced among Habus. In a very special way, they are closer than brothers, more faithful than an old dog, and more dependable than a beating heart. I know of no military relationship that is all-encompassing as the respect, admiration, and friendship that each has for his fellow Habu.

Despite the termination of the SR-71 program, the people who built it, equipped it, maintained it, and flew it have a unique relationship with each other that will never die. Every two years in Reno, Nevada, our "Blackbird Reunion" brings together all the people who have a common bond and fondness for the Blackbird family of aircraft.

On 20 December 1989, Palmdale's SR-71 made a final pass down the Burbank runway

to honor all those who worked in producing the greatest aircraft of the twentieth century. Ben Rich, head of Lockheed's Skunk Works, planned the fly-by and had Kelly Johnson there to watch. Although too sick to get out of the limousine or say anything, Kelly had tears in his eyes as he heard the SR-71 roar past the crowd. Lockheed employee Jim Norris stood there that day and watched the Blackbird pull up and disappear out of sight from the Burbank airport. As Jim watched the aircraft drift out of sight, he recalled the first flight of the SR-71. On that cold December day back in 1964, as the SR-71 became no more than a dark pinpoint, a crew chief standing next to Jim murmured in awe of her beauty and grace, "Her enemies will never be natural." How difficult it must have been back in 1964 to predict that the Blackbird's only real enemies in the end were as natural as you and I.

Appendices

Abbreviations

AB:	Afterburner
AB:	Air Base
AD:	Aerodynamic Disturbance
ADP:	Advanced Development Project
ADS:	Accessory Drive System
AFB:	Air Force Base
AIC:	Air Inlet Computer
AIM:	Air Intercept Missile
ANS:	Astro-Inertial Navigation System
AOA:	Angle of Attack
ARCP:	Air Refueling Control Point
ASARS:	Advanced Synthetic Aperture Radar System
BAK:	Barrier Arresting Kit
BOQ:	Bachelor Officer Quarters
C:	Celsius
CDL:	Common Data Link
CG:	Center of Gravity
CIA:	Central Intelligence Agency
CINC:	Commander in Chief
CINCSAC:	Commander in Chief, SAC
CIP:	Compressor Inlet Pressure
CIT:	Compressor Inlet Temperature
CPA:	Closest Point of Approach
DAFICS:	Digital Automatic Flight and Inlet Control System
DEF:	Defensive System
DIA:	Defense Intelligence Agency
DMZ:	Demilitarized Zone
DOD:	Department of Defense
DO:	Director of Operations
DV:	Distinguished Visitor
EAR:	End Air Refueling
EAS:	Equivalent Air Speed
ECM:	Electronic Counter Measure

EEG:	Electroencephalogram	OBC:	Optical Bar Camera
EGT:	Exhaust Gas Temperature	OER:	Officer Efficiency Report
EKG:	Electrocardiogram	ORI:	Operational Readiness Inspection
ELINT:	Electronic Intelligence	PACAF:	Pacific Air Force
EOB:	Electronic Order of Battle	PARPRO:	Peacetime Aerial Reconnaissance Program
EO:	Electro Optical	PRCSO:	Peacetime Reconnaissance and Certain Sensitive Operations
F:	Fahrenheit		
FAA:	Federal Aviation Agency	PSD:	Physiological Support Division
FCF:	Functional Check Flight	PSI:	Pounds per Square Inch
FOD:	Foreign Object Debris	PVD:	Peripheral Vision Device
FTD:	Field Training Detachment	RAF:	Royal Air Force
FTD:	Foreign Technology Division	RCS:	Radar Cross Section
HF:	High Frequency	RSO:	Reconnaissance Systems Officer
I&W:	Indications and Warning	SAC:	Strategic Air Command
IAS:	Indicated Airspeed	SAM:	Surface-to-Air Missile
ICBM:	Intercontinental Ballistic Missile	SAR:	Special Access Required
IFF:	Identification Friend or Foe	SAS:	Stability Augmentation System
IGV:	Inlet Guide Vane	SECDEF:	Secretary of Defense
IP:	Instructor Pilot	SES:	Shock Expulsion Sensor
JRC:	Joint Reconnaissance Center	SM:	Statute Mile
KEAS:	Knots Equivalent Airspeed	SRC:	Strategic Reconnaissance Center
KIAS:	Knots Indicated Airspeed	SRS:	Strategic Reconnaissance Squadron
LBJ:	Lyndon Baines Johnson	SRW:	Strategic Reconnaissance Wing
mm:	Millimeter	STAN/EVAN:	Standardization and Evaluation
mph:	Miles Per Hour		
NACA:	National Advisory Committee for Aeronautics	TEB:	Triethylborane
		TDI:	Triple Display Indicator
NASA:	National Aeronautics and Space Administration	TDY:	Temporary Duty
		TEOC:	Technical Objective Camera
NCA:	National Command Authority	TS:	Top Secret
NM:	Nautical Mile	UHF:	Ultra High Frequency
NRO:	National Reconnaissance Office	USAF:	United States Air Force
NOFORN:	No Foreign	USAFE:	United States Air Force Europe
NSA:	National Security Agency		

1997 Letter from Senators Levin and Byrd to Secretary of Defense

TED STEVENS, ALASKA, CHAIRMAN

THAD COCHRAN, MISSISSIPPI
ARLEN SPECTER, PENNSYLVANIA
PETE V. DOMENICI, NEW MEXICO
CHRISTOPHER S. BOND, MISSOURI
SLADE GORTON, WASHINGTON
MITCH McCONNELL, KENTUCKY
CONRAD BURNS, MONTANA
RICHARD C. SHELBY, ALABAMA
JUDD GREGG, NEW HAMPSHIRE
ROBERT F. BENNETT, UTAH
BEN NIGHTHORSE CAMPBELL, COLORADO
LARRY CRAIG, IDAHO
LAUCH FAIRCLOTH, NORTH CAROLINA
KAY BAILEY HUTCHISON, TEXAS

ROBERT C. BYRD, WEST VIRGINIA
DANIEL K. INOUYE, HAWAII
ERNEST F. HOLLINGS, SOUTH CAROLINA
PATRICK J. LEAHY, VERMONT
DALE BUMPERS, ARKANSAS
FRANK R. LAUTENBERG, NEW JERSEY
TOM HARKIN, IOWA
BARBARA A. MIKULSKI, MARYLAND
HARRY REID, NEVADA
HERB KOHL, WISCONSIN
PATTY MURRAY, WASHINGTON
BYRON DORGAN, NORTH DAKOTA
BARBARA BOXER, CALIFORNIA

STEVEN J. CORTESE, STAFF DIRECTOR
JAMES H. ENGLISH, MINORITY STAFF DIRECTOR

United States Senate

COMMITTEE ON APPROPRIATIONS
WASHINGTON, DC 20510-6025

October 23, 1997

The Honorable William S. Cohen
Secretary
Department of Defense
The Pentagon
Washington, D.C. 20340

Dear Mr. Secretary:

The recent line item veto of the FY 1998 Defense Appropriations Bill cut all operating funds for the SR-71. It has been brought to my attention that the Air Force may already be moving to place the SR-71 in a status that would prohibit its future use. The SR-71 represents a critical wartime capability and, if properly addressed, a valuable insurance policy on our nation's ability to rapidly expand its imagery gathering capability in the event of a national crisis.

As such, we believe the SR-71 should be placed in flyable storage to be available as the President may direct to meet possible emergency contingencies. Improvements made over the past three years have made this airplane an extremely valuable asset providing real time imagery over a vast area.

Following the Gulf War, General Schwarzkopf testified before Congress that the number one shortfall he encountered was the lack of broad area imagery. We funded the reactivation of the SR-71 to meet that requirement. We continue to witness the intelligence community's attempt to satisfy that same requirement by spending large sums developing Unmanned Aerial Vehicles (UAVs). These UAVs have come at a great expense to the proven manned platforms and have yet to replace the void left by the SR-71.

Would you please review this situation and let us know how you plan to safeguard this valuable asset?

Sincerely yours,

Carl Levin
U.S. Senator

Robert C. Byrd
U.S. Senator

Kelly Johnson's Fourteen Rules of Management

1. The Skunk Works manager must be delegated practically complete control of his program in all aspects. He should report to a division president or higher.

2. Strong but small project offices must be provided both by the military and industry.

3. The number of people having any connection with the project must be restricted in an almost vicious manner. Use a small number of good people (10 percent to 25 percent compared to the so-called normal systems).

4. A very simple drawing and drawing release system with great flexibility for making changes must be provided.

5. There must be a minimum number of reports required, but important work must be recorded thoroughly.

6. There must be a monthly cost review covering not only what has been spent and committed, but also projected costs to the conclusion of the program. Don't have the books ninety days late, and don't surprise the customer with sudden overruns.

7. The contractor must be delegated and must assume more than normal responsibility to get good vendor bids for subcontract on the project. Commercial bid procedures are very often better than military ones.

8. The inspection system as currently used by the Skunk Works, which has been approved by both the Air Force and Navy, meets the intent of existing military requirements and should be used on new projects. Push more basic inspection responsibility back to subcontractors and vendors. Don't duplicate so much inspection.

9. The contractor must be delegated the authority to test his final product in flight. He can and must test it in the initial stages. If he doesn't, he rapidly loses his competency to design other vehicles.

10. The specifications applying to the hardware must be agreed to well in advance of contracting. The Skunk Works practice of having a specification section stating clearly which important military specification items will not knowingly be complied with and reasons therefore is highly recommended.

11. Funding a program must be timely so that the contractor doesn't have to keep running to the bank to support government projects.

12. There must be mutual trust between the military project organization and the contractor with very close cooperation and liaison on a day-to-day basis. This cuts down misunderstanding and correspondence to an absolute minimum.

13. Access by outsiders to the project and its personnel must be strictly controlled by appropriate security measures.

14. Because only a few people will be used in engineering and most other areas, ways must be provided to reward good performance by pay not based on the number of personnel supervised.

Bibliography

Barron, John. *MiG Pilot*. New York: Reader's Digest Association, Inc., 1980.

Byrnes, Donn A., and Kenneth D. Hurley. *Blackbird Rising*. Los Lunas, NM: Sage Mesa Publications, 1999.

Crickmore, Paul. *Lockheed Blackbird, Beyond the Secret Missions*. Oxford: Osprey Publishing Ltd., 2010.

Goodall, James. *SR-71 Pilot's Manual*. Minneapolis: Zenith Press, 1992.

Graham, Richard H. *SR-71 Revealed: The Inside Story*. Minneapolis: Zenith Press, 1996.

———. *SR-71 Blackbird: Stories, Tales, and Legends*. Minneapolis: Zenith Press, 2002.

———. *Flying the SR-71 Blackbird: In the Cockpit on a Secret Operational Mission*. Minneapolis: Zenith Press, 2008.

Jenkins, Dennis R. *Lockheed SR-71/YF-12 Blackbirds*. North Branch, MN: Specialty Press Publishers and Wholesalers, 1997.

Miller, Jay. *Lockheed's Skunk Works*. Tulsa, OK: Aerofax, 1995.

Remak, Jeannette, and Joseph Ventolo Jr. *A-12 Blackbird Declassified*. Minneapolis: Zenith Press, 2001.

Rich, Ben, and Leo Janis. *The Skunk Works*. New York: Little, Brown and Company, 1994.

Robarge, David. *Archangel: CIA's Supersonic A-12 Reconnaissance Aircraft*. Washington, D.C.: Central Intelligence Agency, 2007.

Index